Supplement to Fluid Mechanics, Water Hammer, Dynamic Stresses, and Piping Design

This 2015 Supplement includes recommended additions and corrections to the original text, which was published by ASME Press in 2013

Robert A. Leishear, PhD, P. E.

This manuscript has been authored by Savannah River Nuclear Solutions, LLC under Contract No. DE-AC09-08SR22470 with the U.S. Department of Energy. The United States Government retains and publisher, by accepting this article for publication, acknowledges that the United States Government retains a non-exclusive, paid-up, irrevocable, worldwide license to publish or reproduce the published form of this work, or allow others to do so, for United States Government purposes.

© 2015, The American Society of Mechanical Engineers (ASME), 2 Park Avenue, New York, NY 10016, USA (www.asme.org)

All rights reserved. Printed in the United States of America. Except as permitted under the United States Copyright Act of 1976, no part of this publication may be reproduced or distributed in any form or by any means, or stored in a database or retrieval system, without the prior written permission of the publisher.

INFORMATION CONTAINED IN THIS WORK HAS BEEN OBTAINED BY THE AMERICAN SOCIETY OF MECHANICAL ENGINEERS FROM SOURCES BELIEVED TO BE RELIABLE. HOWEVER, NEITHER ASME NOR ITS AUTHORS OR EDITORS GUARANTEE THE ACCURACY OR COMPLETENESS OF ANY INFORMATION PUBLISHED IN THIS WORK. NEITHER ASME NOR ITS AUTHORS AND EDITORS SHALL BE RESPONSIBLE FOR ANY ERRORS, OMISSIONS, OR DAMAGES ARISING OUT OF THE USE OF THIS INFORMATION. THE WORK IS PUBLISHED WITH THE UNDERSTANDING THAT ASME AND ITS AUTHORS AND EDITORS ARE SUPPLYING INFORMATION BUT ARE NOT ATTEMPTING TO RENDER ENGINEERING OR OTHER PROFESSIONAL SERVICES. IF SUCH ENGINEERING OR PROFESSIONAL SERVICES ARE REQUIRED, THE ASSISTANCE OF AN APPROPRIATE PROFESSIONAL SHOULD BE SOUGHT.

ASME shall not be responsible for statements or opinions advanced in papers or . . . printed in its publications (B7.1.3). Statement from the Bylaws.

For authorization to photocopy material for internal or personal use under those circumstances not falling within the fair use provisions of the Copyright Act, contact the Copyright Clearance Center (CCC), 222 Rosewood Drive, Danvers, MA 01923, tel: 978-750-8400, www.copyright.com.

Requests for special permission or bulk reproduction should be addressed to the ASME Publishing Department, or submitted online at https://www.asme.org/shop/books/book-proposals/permissions

ASME Press books are available at special quantity discounts to use as premiums or for use in corporate training programs. For more information, contact Special Sales at customercare@asme.org

Library of Congress Cataloging-in-Publication Data

Library of Congress has already cataloged the main work as:
Leishear, Robert Allan.
Fluid mechanics, water hammer, dynamic stresses, and piping design / Robert A. Leishear.
p. cm.
Includes bibliographical references and index.
ISBN 978-0-7918-5996-4
1. Fluid mechanics. 2. Piping—Design and construction. 3. Water hammer.
I. Title
QC145.2.L45 2012
660'.283–dc23
2012016745

On the cover: Water hammer knocked 2000 feet of 24 inch diameter piping to the ground during a 1984 fluid transient accident at Savannah River Site in South Carolina. (Courtesy of Savannah River National Laboratory)

To Janet Leishear who made this work possible.

Table of Contents

I. Introduction	9
II. Appendix B: Summary of Piping Design	
II.A. Appendix B.1 – A Discussion of DLFs for Piping Design	11
1. Introduction	12
2. Piping failures due to water hammer (fluid transients)	13
3. DLF's	15
4. DLF's for bending stresses	16
5. DLF's for valves and short pipes	19
5.1 Pressure vessels	19
5.2 Valves	19
5.2.1 Valve explosion example	19
5.2.2 Valve leak example	19
5.3 DLF's for short pipes	21
6. DLF's for long pipes	25
6.1 Flexural resonance theory	25
6.2 Dynamic stress theory	25
6.3 Tests in gas filled piping	25
6.4 Tests in water filled piping	28
6.5 Comparison of DLF theories for long pipes	29
7. SRS piping failures due to water hammer	33
7.1 Bending and hoop stress piping failures in 6 and 8 inch diameter, ductile iron piping	33
7.2 Hoop stress piping failures in 2 inch diameter, carbon steel piping	34
8. Plastic deformations and pipeline explosions	38
9. DLF's for plastic deformations	39
II.B. Appendix B.2 – Design of Piping Systems for Dynamic Loads from Fluid Transients	45
Foreword	46
Introduction	47
1. Purpose	48
1.1 Scope	48
1.2 Terms and definitions	48
2. Materials	50
3. Classes of fluid transients	51
4. Fluid transients in liquid filled piping systems	53
4.1 Transients due to valve openings and closures	55
4.1.1 Pressure surges due to sudden valve openings and closures	55
4.1.2 Pressure wave velocities in liquid filled piping	59
4.1.3 Reflected and transmitted waves in liquid filled piping	62

	4.2 Transients during centrifugal pump start-up and shut-down	66
	4.2.1 Centrifugal pump operations	66
	4.2.2 Operations of centrifugal pumps in parallel	68
	4.3 Operations of positive displacement pumps	68
5.	**Fluid transients in liquid-gas filled piping systems**	**69**
	5.1 Reduction in transient effects due to trapped air pockets	69
	5.2 A comparison between sudden valve closure, slow closure, and air pocket effects	69
	5.3 Transient effects due to large air quantities in liquid filled piping systems	69
	5.3.1 Transient effects due to liquid filling of air filled piping systems	71
6.	**Fluid transients in liquid-vapor filled piping systems**	**72**
	6.1 Pressure surges due to void formation and vapor collapse	72
	6.2 Fluid transients due to vapor collapse in liquid filled piping systems	72
	6.2.1 Void formation and vapor collapse	73
	6.2.2 Damage mechanisms and corrective actions for vapor collapse in liquid filled systems	76
	6.3 Condensate induced water hammer	76
	6.3.1 Condensate induced water hammer in horizontal pipes	76
	6.3.2 Water cannon	77
	6.3.3 Damages and corrective actions for condensate induced water hammer	78
	6.4 Slug flow	79
7.	**Fluid transients in vapor or gas filled piping systems**	**81**
	7.1 Pressure loads on safety valve installations	81
	7.2 Reaction forces on open discharge pipes	81
	7.2.1 Forces on discharge elbows	81
	7.2.2 Forces on vent pipes	82
8.	**Piping system response: loads, stresses, and reactions**	**84**
	8.1 Piping system loads	84
	8.2 DLF's and dynamic pipe stresses	84
	8.3 Dynamic hoop stresses	87
	8.3.1 DLF's for hoop stresses	87
	8.3.2 Reflected pressure waves and hoop stresses	88
	8.4 Dynamic bending forces and reactions	88
	8.4.1 Bending forces and DLF's	89
	8.4.2 Reactions	91
	8.5 Loads on other piping components	91
9.	**Damage assessment**	**93**
10.	**References**	**97**
11.	**Nomenclature**	**98**

III. Updates, Revisions and Corrections to this Book	100
III.A. Uncertainty Analysis	100
Measurement uncertainty	100
Process variations	104
III.B. Vibration Analysis	104
III.C. Corrections	109
Index	117

I. INTRODUCTION

Numerous reviews were performed on the original text for this book, and manuscripts for this book were used to teach the topics in the text to four separate classes of engineers at SRS prior to publication. Comments from these four day classes were incorporated into the manuscripts between classes to improve the quality of presentation. Even so, improvements and some corrections to the text have been noted since ASME classes were taught after publication. This Supplement captures these additions and changes. Also included in this Supplement are appendices to summarize this book through "A Discussion of DLF's for Piping Design" and the "Design of Piping Systems for Dynamic Loads From Fluid Transients".

The format of this Supplement needs to be related to the format of this book:

- Chapter numbers are supplied in this Supplement to be consistent with chapter numbers in this book.
- Figure numbers are selected to be consistent with figure numbers in this book. For example, Figure 5.22A from this Supplement is required between Figures 5.22 and 5.23 in this book.
- Equations are treated similarly. As examples, the equation between 2.101 and 2.102 is numbered as 2.101A.
- New references are provided in this Supplement in their entirety without any attempt to sequence them with the references in this book. When references already listed in this book are used, they are simply referred to as they are in this book. For example, Coleman and Steele [184] are listed as Reference 184 in the References at the end of this book.
- New symbols and abbreviations will be defined in this Supplement as required, and will not be listed in the nomenclature of Appendix A.3 at the end of this book.

II. Appendix B: Summary of Piping Design

II.A. Appendix B.1: A Discussion of DLF's for Piping Design

R. A. Leishear, PhD, P. E.

1. Introduction

This book provides a comprehensive discussion of DLF's and their applications to piping design as presently understood ("Fluid Mechanics, Water Hammer, Dynamic Stresses, and Piping Design", R. A. Leishear, ASME Press, 2013). A summary is provided here, and figures in the following discussion are copied from my book unless otherwise noted. All in all, this summary focuses the lengthy topic of dynamic load factors into a shorter, more manageable, discussion of dynamic load factors. More research should be performed, since there is yet much to be learned in this area of research to prevent continuing multi-million dollar damages and prevent accidental deaths.

My goal for this book, quite simply, is to teach others what I have learned over the past twenty years of study. Together we are a better engineer.

2. Piping failures due to water hammer (fluid transients)

Let me first make a few general observations that are supported by fact.

1. Hundreds of water hammer accidents have been documented in the literature, as discussed by EPRI and the NRC (referenced in my book).
2. Brittle fractures of valves and piping have been observed where they have exploded due to water hammer. See the discussions in my book for the DOE, Hanford fatality and a New York City pipeline explosion.
3. Plastic failures have occurred due to explosions in piping, where eight inch diameter steel pipes were shredded in nuclear facilities.
4. One of the most widely reported water hammer accidents in recent years was cited at a Russian hydroelectric plant. This accident was later attributed to fatigue failures of turbine mounting bolts. Water hammer may have affected other turbine failures, but was not investigated. All failures are not caused by water hammer, and designers and engineers need to be careful how to apply theory.

Now, let me make a few general observations that appear to be correct, but have yet to be universally proven.

1. Significant water hammer damages have occurred due to plastic deformation. Piping has been bent and knocked from its piping supports.
2. Most pipe ruptures occur in brittle components. Few have apparently been caused due to plastic deformations caused by water hammer in piping (www.kirsner.org). Additionally of note, J. Frey presented piping failure examples in one of his classes on B31.1 piping, where piping was believed to fail due to thermal fatigue. Perhaps water hammer was an influential factor. He also noted that piping fractures occurred at dead ends, and encouraged the removal of dead end piping (dead-headed) from service. Perhaps these failures are indicators of reflected pressure

waves in piping. I certainly am not second guessing Mr. Frey, but we continue to learn new information in this field.
3. Usually, valve leaks are observed in systems before fatigue fractures occur.
4. Although relief valves are too slow to actuate fast enough to prevent pressure waves from traveling throughout a system, relief valves will open shortly after the system is pressurized due to water hammer.
5. Several times a year major pipeline ruptures are reported in the press. According to the Los Angeles Times (August 2014), approximately 244,000 piping failures have occurred due to aging in U. S. municipal water systems as reported by AWWA. I suspect, but have not proven, that many of these failures were in fact caused by water hammer. Many of the piping failures that I have investigated were reported as aging failures, but water hammer was shown to be the failure cause. I do not believe that materials fail because they get old. They fail in fatigue or due to corrosion. For cold water systems, one likely cause of fatigue is water hammer. When someone says that a pipe cracked because it was old, fatigue is usually the failure cause, and water hammer is frequently the fatigue cause for piping. There is no failure mechanism referred to as "old".
6. In many, if not all, cases of water hammer damages, reducing the fluid transient pressure surges may reduce the stresses on existing cracks to stresses low enough to prevent those cracks from growing any further.

3. DLF's

A brief summary of DLF's follows for pipes subjected to suddenly applied, constant pressures (step pressure changes). As the pressures are more slowly applied, the DLF's will reduce to 1

1. The DLF equals the ratio of the maximum stress occurring during a vibration cycle divided by the static stress that would be incurred if an identical load was slowly applied.
2. The maximum DLF < 2 for elastic bending stresses, but most bending stress calculations should be performed using computer models. This value may be exceeded when opposing bends are present in a piping system. Computer calculations are preferred for complex piping geometries.
3. DLF's for elastic hoop deformations are applicable to fatigue failures. The fatigue limit for many common piping materials is linear up to nearly the yield point of those materials, and simplified equations may be used to model linear material responses.
4. The maximum DLF < 2 for valves, short pipes and components.
5. The maximum DLF < 4 for hoop stresses in longer pipes. Although not yet proven, a long pipe seems to be described as pipes longer than 5 to 20 pipe diameters, or so.
6. The DLF's for plastic deformation are not presently understood, but are estimated to be near one due to excessive damping during plastic deformation.
7. DLF's for hoop stresses and components may be nearly doubled near the ends of pipelines, where reflected pressure waves double the pressure surges in the piping.

4. DLF's for bending stresses

For bending stresses, only piping systems that can be represented by a single frequency response can be readily adapted to the use of DLF's to determine a maximum stress. Presently, a figure from B31.1 has been adapted for this Appendix (Figure 8.4B). The figure was further simplified by straight line approximations for B31.1 to express the impulse force shown in Figure 7.8. In B31.1, this figure applies to piping with a single elbow connected to a relief valve at one end, where the other end is unattached and free to move. For this Appendix, this figure can also be applied to piping with a single elbow but fixed at both ends.

The curve was generated from the equations of vibration for single degree of freedom (SDOF) oscillators, which were in turn derived from Newton's Law. These equations also assume that the piping material responds in a linear manner according to Hooke's Law for elastic materials, and damping is neglected. Chapter 7 of my text provides a detailed discussion and further references for the derivations of SDOF responses (Biggs, Harris, etc.). A DLF may be determined for elastic bending by assuming that the input represents the water hammer wave, where the input may be a constant pressure step (DLF = 2 at $t_o / \tau = 0$), or any other

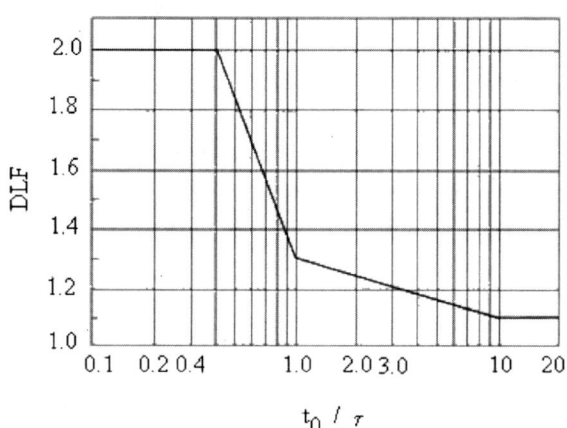

From Appendix B.2: Figure 8.4B Effects of short duration pressure surges on system response (stresses or reactions) [Ref. 6].

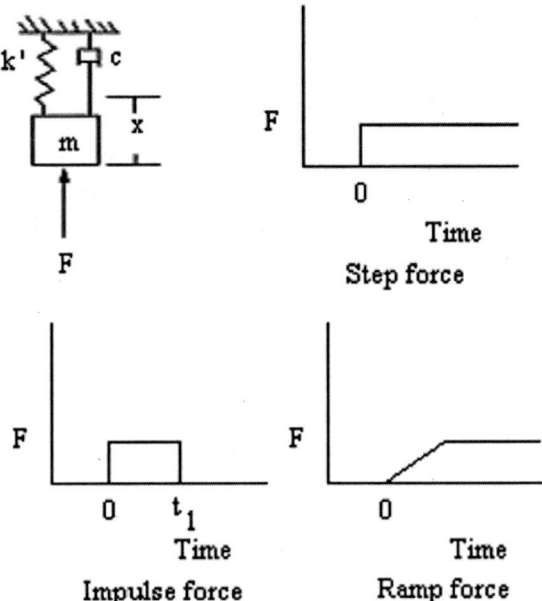

Figure 7.8 SDOF excitations/applied forces.

impulse pressure surge such that the application of load, t_o, is divided by the vibration period, τ, of the piping to find the DLF. The DLF is then multiplied times the maximum calculated static stress in the piping to obtain the maximum dynamic stress. Note that even though damping will reduce dynamic effects during localized plasticity, elastic follow-up will result in DLF's near two during bending.

For more complex piping arrangements, computer simulations are recommended, which do not require DLF's since the dynamic effects of water hammer loads are presented within the computer simulation. Even so, the assumed stiffness at elbows, etc. will affect the predicted stresses. Further research in this area along with some experimental research is recommended to better understand the accuracy of computer modeling for dynamic bending.

In many piping failures, bending stresses dwarf hoop stresses as shown in Figures 9.4 and 9.7, where an example of a water hammer induced step pressure and resultant stresses are depicted.

18 Appendix B.1

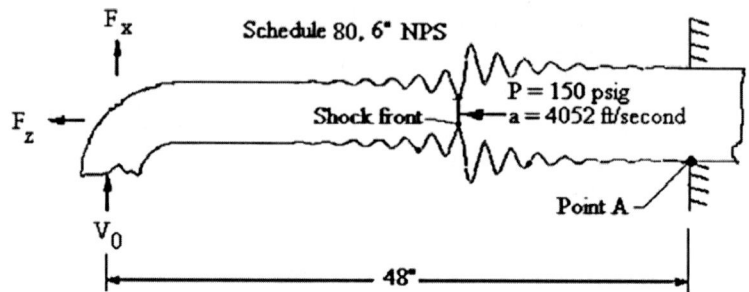

Figure 9.4 Shock impingement on an elbow.

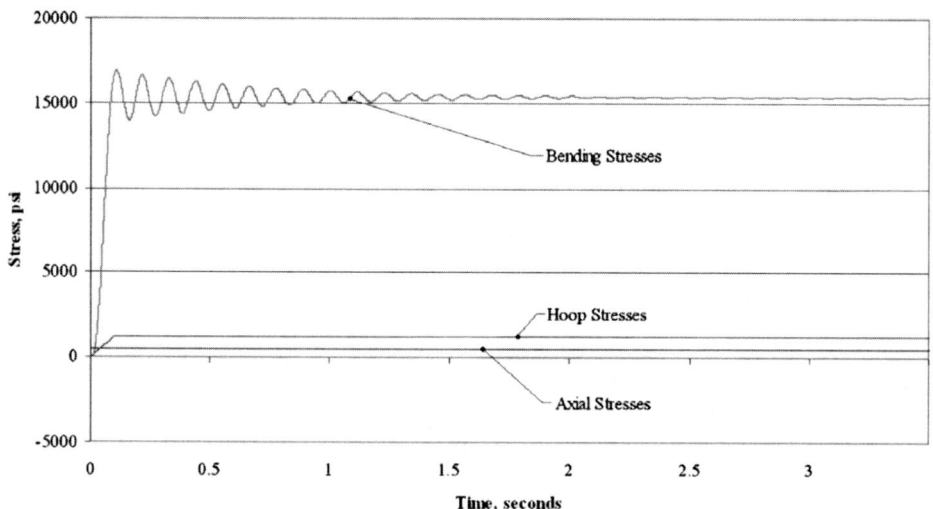

Figure 9.7 Stress response for a 0.1-second rise time.

5. DLF's for valves and short pipes

The DLF < 2 for valves and short pipes. A first approximation is that any valve or length of pipe subjected to a step change in pressure will have a DLF less than or equal to 2. If damping is neglected, DLF = 2 for a constantly applied pressure. Figure 8.4B is also applicable to hoop stresses, but hoop stress frequencies are typically so high that few cases occur where short load application times, t_o, are applicable to water hammer loading. Even so, damping yields a DLF < 2.

5.1 Pressure vessels

Some of the earlier experimental DLF work was performed on pressure vessels. A maximum DLF of 2.5 was determined through strain gauge measurements during internal blast loading in a 20 foot long, 2 foot diameter pressure vessel (Malherbee [234], 1966). There is more recent experimental and theoretical research on explosions inside pressure vessels, which is not considered here. In fact, Code Case 2564 was issued to consider internal explosions in vessels and design requirements for Section VIII pressure vessels.

5.2 Valves

5.2.1 Valve explosion example

With respect to valves, one of the most well documented water hammer induced valve explosions occurred at a DOE facility in Hanford, Washington. An operator was killed from steam inhalation inside the valve pit shown in Figure 6.4. A valve was opened to mix 55°F condensate with 110 psig steam to accidentally cause water hammer. A thorough investigation was performed which included computer simulations of the fluid flow, hydrostatic bursting of similar valves, FEA valve modeling, experimental simulation of condensate induced water hammer, and electron microscopy of failed valve surfaces. Among other conclusions, the report stated that there was a DLF of at least 1.7 to cause this valve explosion. Although not stated in their report, the DLF may have been higher due to reflected wave effects. Their FEA models were confined to static loading of the applied water hammer pressure.

5.2.2 Valve leak example

Another example of valve failures due to water hammer occurred at SRS, where I was working as the Shift Engineer responsible for

20 Appendix B.1

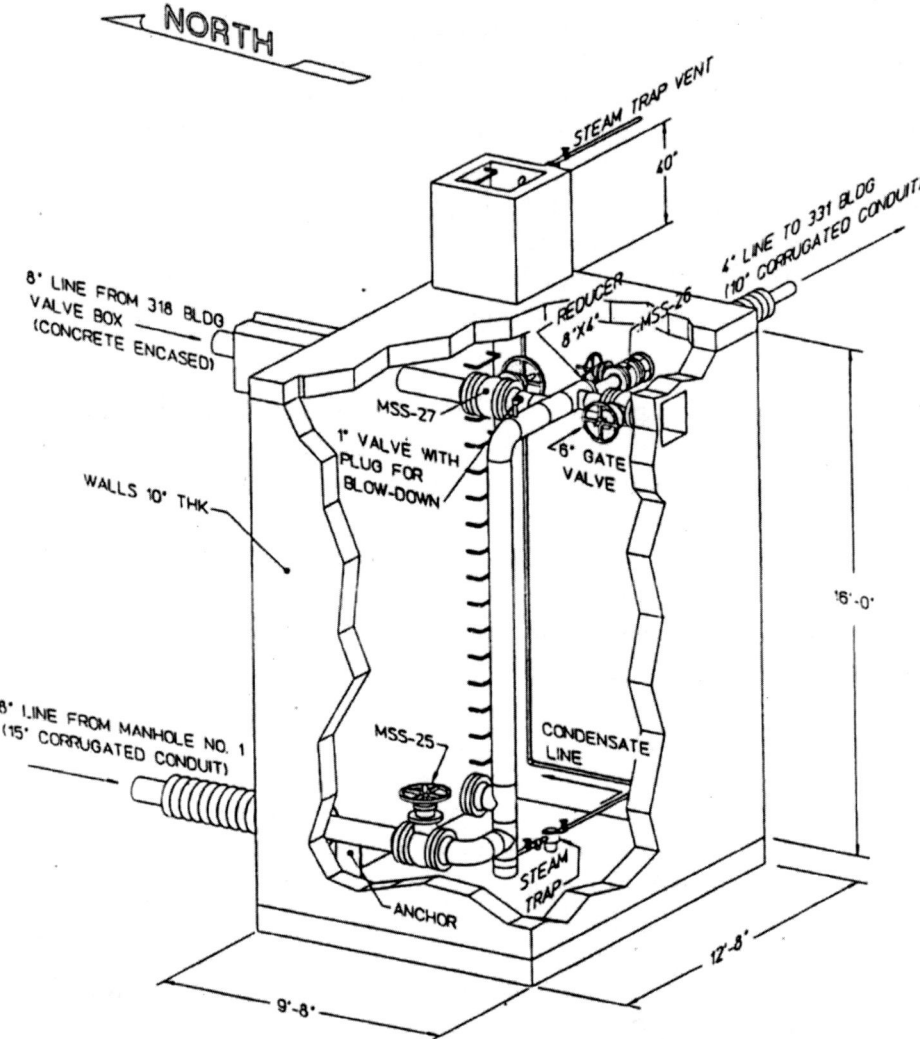

Figure 6.4 Hanford accident piping configuration (Green [195]).

Operations oversight on the shift immediately following water hammer damages, which occurred at shift change. I interviewed all operators who manipulated valves at the time of the incident, and performed a series of water hammer models to determine maximum pressures for each valve operation, as shown in Figure 5.33. A typical leak is shown in Figure 5.30.

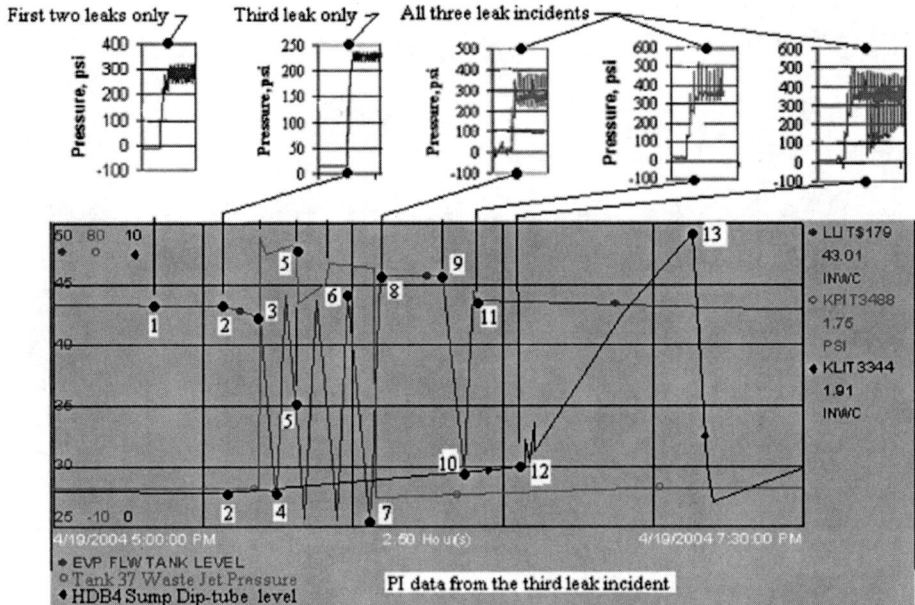

Figure 5.33 Valve leaks and pressure calculations.

Piping was not damaged, but numerous valves leaked. In non-nuclear facilities, leaks may not always be a major concern, but at SRS these valves controlled the flow of radioactive liquid nuclear waste, and repairs for this one incident exceeded a quarter of a million dollars. Water hammer induced pressures were calculated to be well above design, but the valves were shown to withstand the applied pressures unless a dynamic load factor of nearly 2 was applied to the static pressure calculation. Further investigation determined that similar failures had occurred at SRS for many years, and valves were frequently rebuilt using specialized procedures to improve their ability to withstand leaks. Water hammer was previously unrecognized as the cause of valve failures, and the valve design was considered to be at fault. When the hammering was eliminated, no more valve leaks occurred over the past ten years. Water hammer was the problem all along.

5.3 DLF's for short pipes

Some of my graduate research focused on DLF's in short pipes, using FEA models for 14 inch long, 8 inch diameter steel pipes. One of the conclusions was that the DLF equaled 2 at some cross section on the piping regardless of material or end constraints, in the absence of damping.

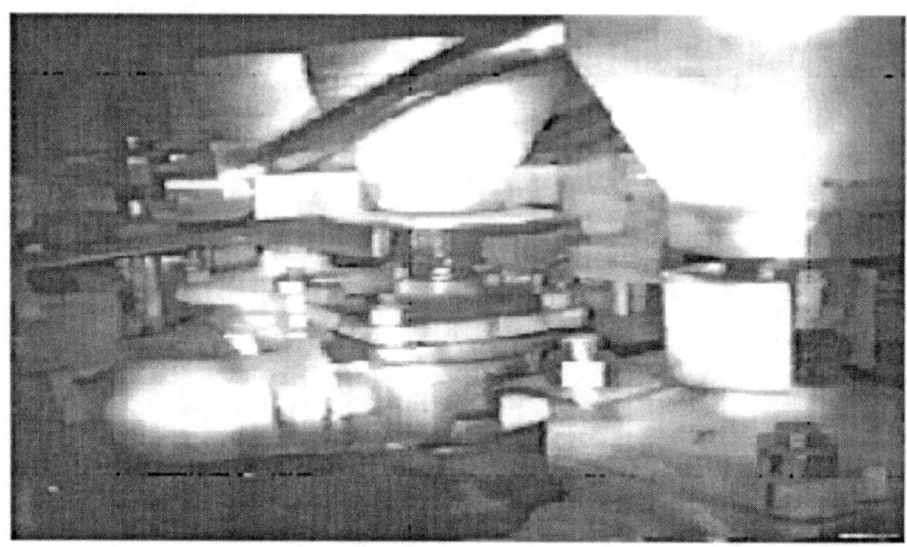

Figure 5.30 Remote photo of a packing leak.

That is, the SDOF approximation was applicable for at least one location along the pipe, circling the diameter of the pipe. FEA models for two fixed ends, one fixed and one free end, one fixed and one sliding end were all modeled for different step pressures and pipe wall materials. In the elastic range of the piping materials, the maximum DLF equaled 2 for all models. For a steel pipe, the results are shown in Figures 8.9 and 8.16, where the maximum stress occurred at the maximum deflection on the inner pipe wall.

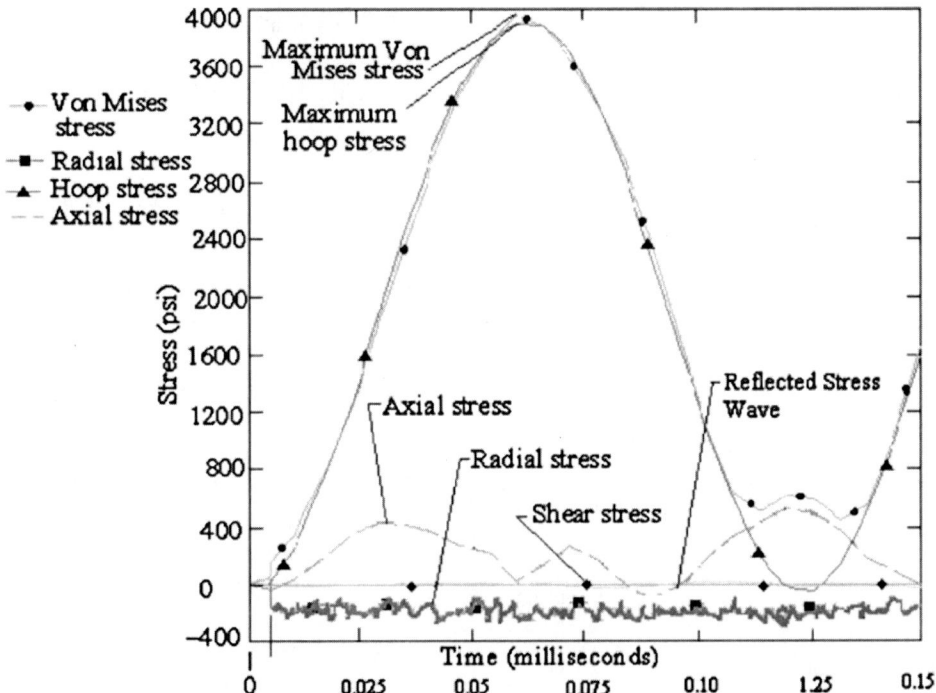

Figure 8.9 FEA, Calculated maximum stresses in pipe with both end fixed.

24 Appendix B.1

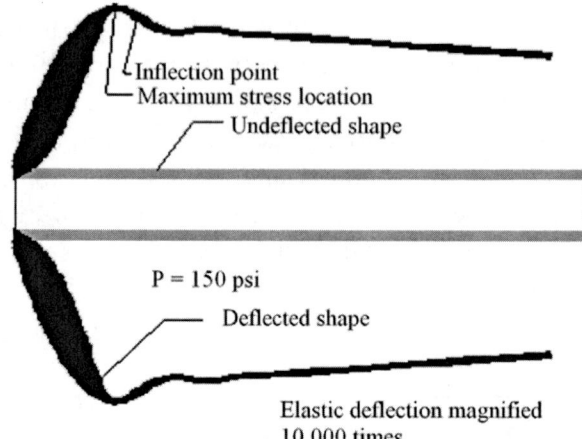

Figure 8.16 Deflection in a free-end pipe, one end fixed.

6. DLF's for long pipes

One might assume that the DLF equals two for any length of pipe, but this assumption is incorrect. In some of my earlier publications, I was also of this incorrect opinion. Experimental facts convinced me otherwise, where research from three researchers, including myself, is available in the literature. Experimentally, a DLF < 4 exists where a precursor wave and an aftershock wave are present near a shock wave moving along the bore of a pipe.

6.1 Flexural resonance theory

Simkins, Beltman and Shepherd all applied a theory to pressure waves in tubes referred to as the flexural resonance theory, after Tang, 1963 (Paragraph 8.4). Simkins performed tests on artillery to measure the strains due to nearly constant pressures behind a shell exiting the steel barrels of cannons. Beltman and Shepherd performed experiments on step pressure waves in gas filled 6 inch diameter Aluminum tubes. Each of these researchers obtained results that indicated that there was a critical velocity, and that near that velocity the DLF was slightly less than 4. Even so, there models assumed an infinite stress in the pipe at the shock wave, which is inconsistent with the DLF equal to 4. The following discussion of dynamic stress theory therefore supersedes flexural resonance theory.

6.2 Dynamic stress theory

I developed a theory based on the assumptions that the precursor wave has a maximum DLF = 2 and the aftershock wave also has a maximum DLF = 2 (Paragraph 8.5). Each wave is expressed as an elastic vibration that responds to a step increase in pressure. As the wave moves, the precursor and aftershock waves add to yield a vibration behind the pressure wave with a maximum DLF equal to 4.

6.3 Tests in gas filled piping

This theory reasonably predicts the maximum stresses measured in Beltman's tests for 6 inch diameter, 6061 aluminum tubing (Figures 8.25, 8.28 and 8.38), as well as stresses measured in additional tests that I performed at SRS.

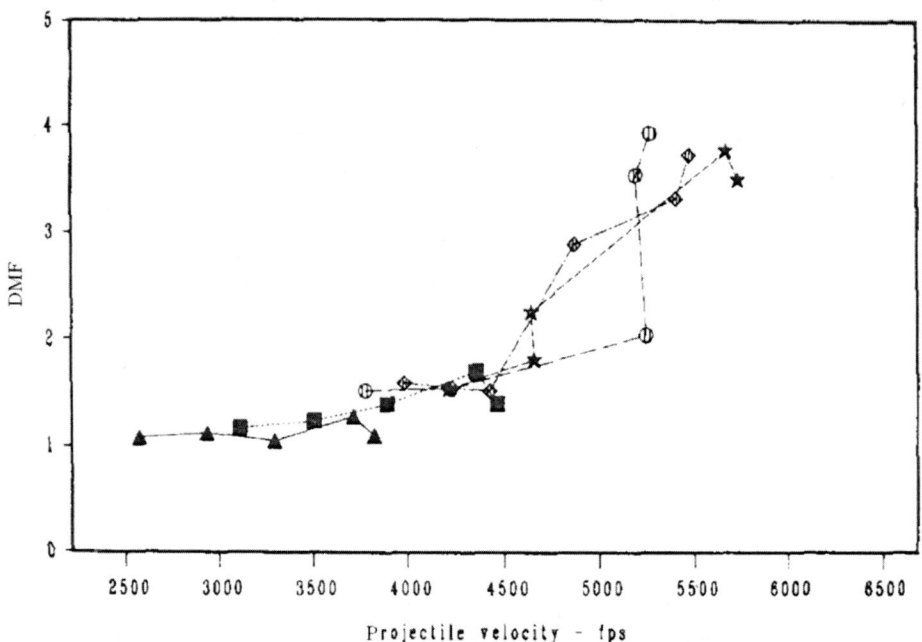

Figure 8.25 Hoop stress test results for a 120-mm gun tube (Simkins [227 and 228]).

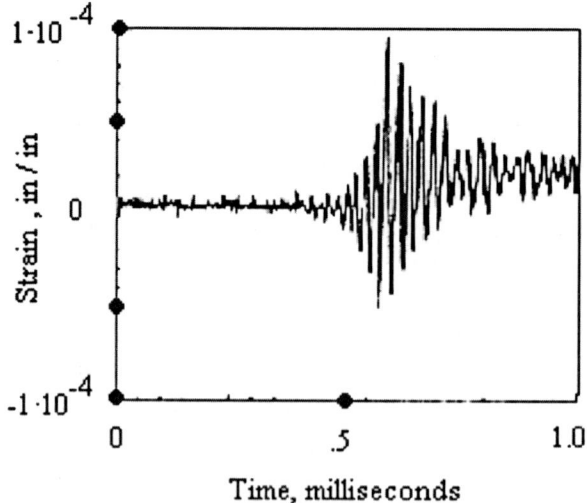

Figure 8.28 Strains at a point on the outer aluminum pipe wall, shock velocity = 3175 ft/second (Beltman et al. [229]).

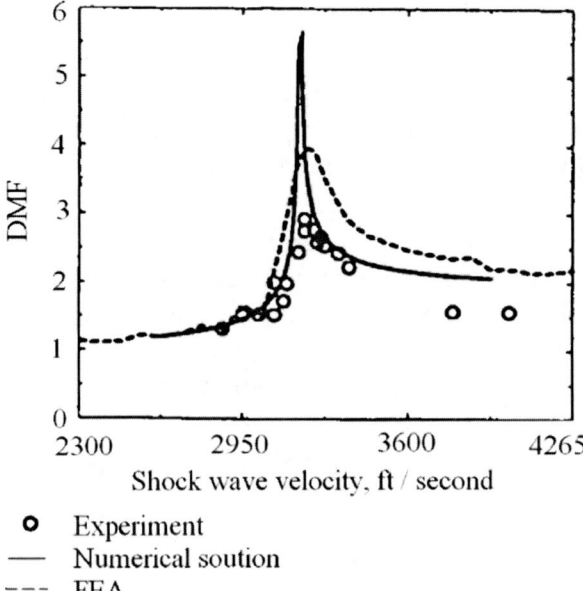

o Experiment
— Numerical soution
--- FEA

Figure 8.29 Dynamic amplification data for hoop stresses in an aluminum tube (Beltman et al. [229]).

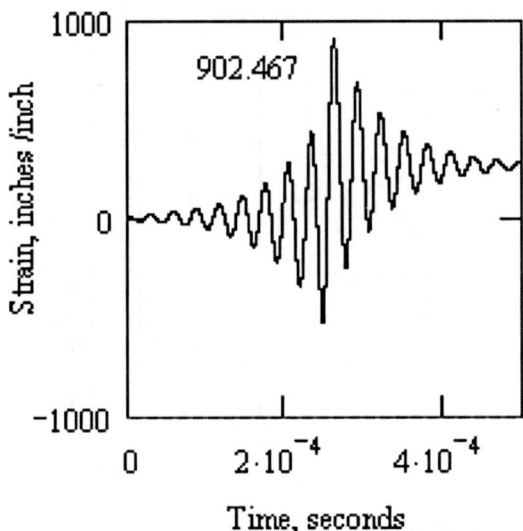

Figure 8.38 Maximum hoop stress at the outer aluminum wall due to a shock.

6.4 Tests in water filled piping

For the SRS hoop stress testing, an in-service 2 inch diameter piping system was instrumented to measure the pressure in the piping as well the strains at numerous points along the piping. This case study is considered throughout the book in Examples 2.13, 4.9, and 8.8. The system setup is shown in Figures 2.88 and 8.41. A typical pressure is shown in Figure 8.43, and a series of strain waves are shown in Figures 8.47, 8.48 and 8.49. The pressure wave travelled along bore of the pipe at a wave speed of approximately 2680 feet/second, which was less than predicted. However, the maximum experimental DLF equaled 3.78, which was expected to be slightly less than 4. To incorporate damping, DLF estimates were approximated to lie between 3.51 and 3.96, using Equation 8.138. Note that damping is outside the scope of

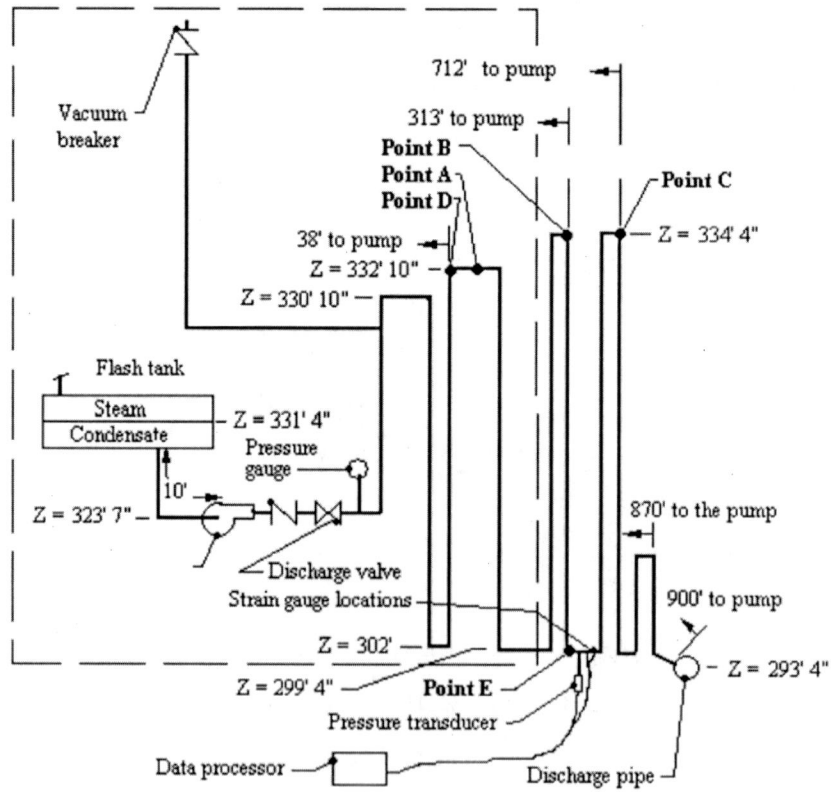

Figure 2.88 Description of a condensate system.

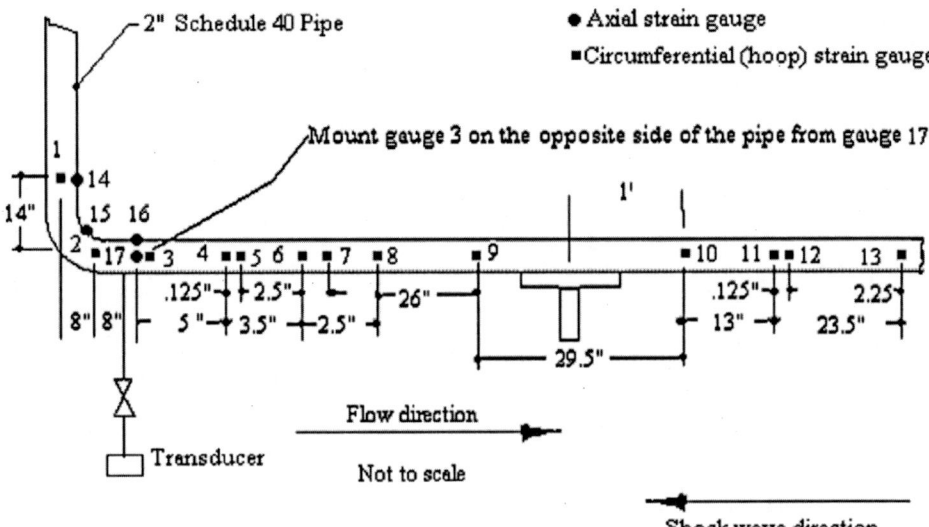

Figure 8.41 Test setup for water hammer strain evaluation at SRS.

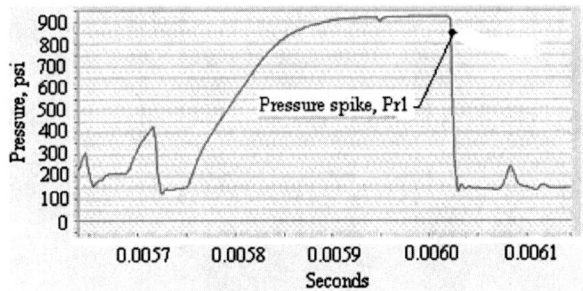

Figure 8.43 Pressure data (zoomed in on Pr1).

Appendix B.2, but damping was considered to evaluate the applicability of the dynamic stress theory with respect to experimental data.

Although the maximum stresses are well predicted using the dynamic stress theory, the frequency is not very well predicted for Schedule 40 piping, which is a result similar to flexural resonance theory predictions. Also, the maximum DLF may decrease below a value of 4 for larger diameter piping, but further research is required.

6.5 Comparison of DLF theories for long pipes

All of the basic theories assume thin wall approximations, since thick wall approximations have not been developed. FEA models or

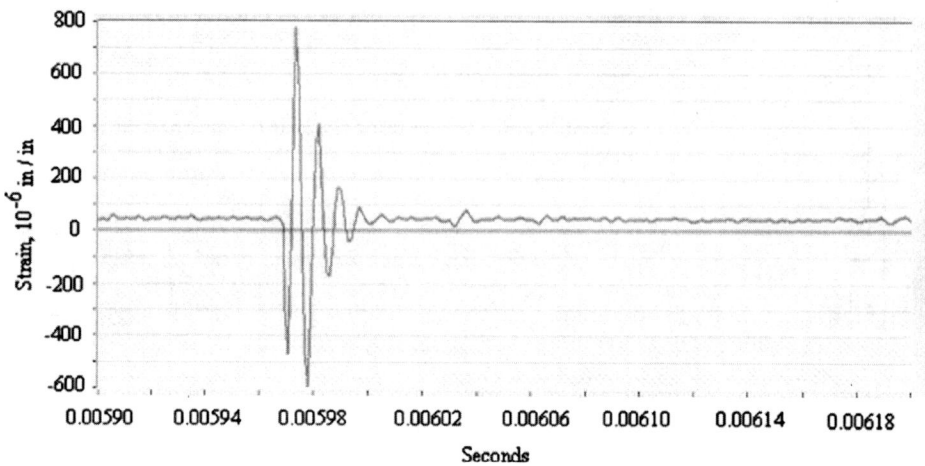

Figure 8.47 Typical hoop strain measurements at point 12.

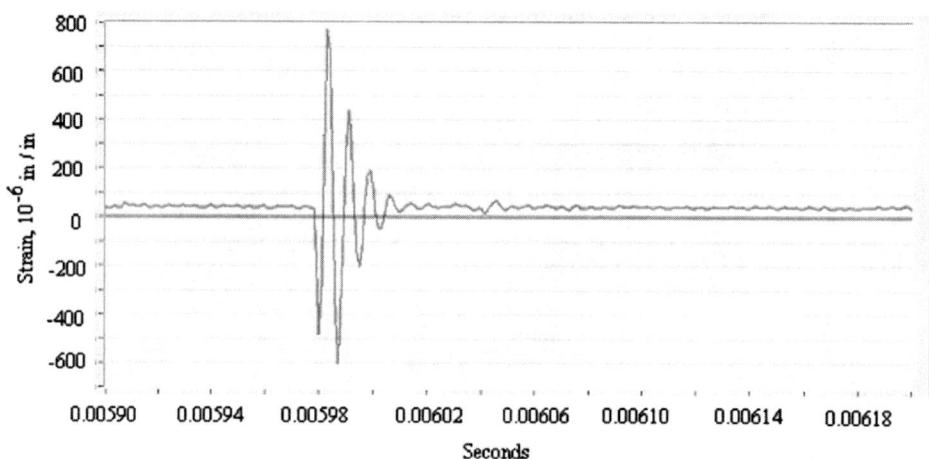

Figure 8.48 Typical hoop strain measurements at point 11.

experiments are required to fully understand DLF's. A brief summary of the presented theories follows:

1. All closed form theories incorrectly predicted the experimental frequency of hoop stress vibrations for a 2 inch pipe, but the maximum hoop stress values are nearly independent of frequency.
2. Flexural resonance theory yields a DLF = 1.1 for hoop stresses in 2 inch schedule 40 pipe (Table 8.1 in the book), where

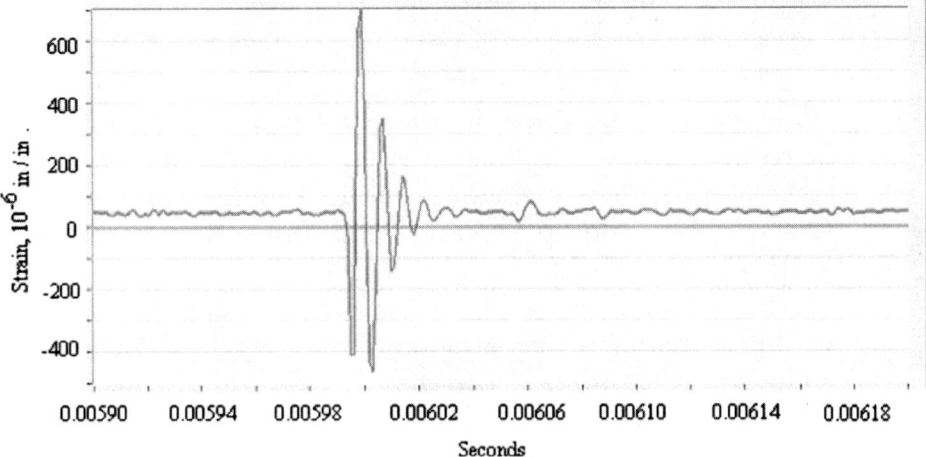

Figure 8.49 Typical hoop strain measurements at point 10.

Table 8.1 DMF for flexural resonance in fixed-end pipes.

Sch 40, NPS	V_{cr}, ft/second, Eq. (8.61)	Water hammer wave speed, a, ft/second, Eq. (5.20)	Ratio of speeds, a/V_{cr}, Fig. 8.22	DMF, Fig. 8.21
1	15,383	4625	0.30	1.1
2	9067	4515	0.50	1.1
3	8469	4499	0.51	1.1
4	7353	4483	0.61	1.2
6	5140	4355	0.85	1.3
8	4245	4299	1.01	4.5
10	3909	4252	1.09	3.3
12	3118	4176	1.34	2.2
14	3216	4210	1.31	2.2
16	2806	4052	1.44	2.1
Sch. 5S, NPS				
1	6401	4419	0.69	1.2
2	3008	4136	1.37	2.2

experiments showed a DLF near 4. Experiments have never yielded a DLF > 4, as predicted by flexural resonance theory.
3. The dynamic stress theory predicts that DLF = 4 in the absence of damping. If critical velocity effects are assumed, the DLF may be reduced below 4 at different points along the pipe wall. If a critical velocity is not assumed, the maximum DLF = 4. The DLF < 4 always due to damping, but damping is not considered in the Appendix B.2.
4. Flexural beam theory yields a comparable maximum DLF = 4, by differentiating the equation of motion for the defections of a pipe wall. According to this theory, there will be a maximum DLF = 4 at the critical velocity.
5. Experimental maximum DLF's of 4 were observed in 2 inch pipes, 6 inch tubes, and 2.36 (60 mm) and 4.72 inch (120 mm) gun barrels.
6. Regardless of theory, the maximum DLF of approximately 4 occurred at water hammer wave speeds in 2 inch Schedule 40 pipes and 6 inch diameter tubes.

7. SRS piping failures due to water hammer

Hundreds of piping failures occurred at SRS over a forty year period in a cooling water system used to cool nuclear waste storage tanks, which are self-heated due to radioactivity within the stored liquid waste. In the 1970's, a study concluded that corrosion was a probable cause of piping failures in tanks constructed in the 1950's. When water hammer was eliminated, piping failures stopped. The only new pipe failures since 2002 occurred when water hammer protective measures were inadvertently removed from service. Corrosion was not the failure cause. The piping was part of a B31.3 designed system used to cool radioactive liquid waste in million gallon tanks.

Hundreds of failures occurred in 2 inch diameter, Schedule 40, carbon steel cooling coils and 6 and 8 inch diameter ductile iron piping. The cooling water inside the piping was treated with a sodium chromate corrosion inhibitor and the waste is treated with a sodium hydroxide corrosion inhibitor. Some of the cooling coils extended above the treated waste, which led to the conclusion that corrosion was a likely failure cause.

This case study was considered throughout my book as a demonstration of the application of DLF's, using Examples 2.13, 4.9, 5.6, 5.22, 5.23, 5.6, and 9.6.

SRS has also lost other piping systems due to fatigue, where water hammer was a probable cause.

7.1 Bending and hoop stress piping failures in 6 and 8 inch diameter, ductile iron piping

A typical model for an underground piping fatigue failure analysis is shown in Figure 9.13. Autopipe® was used to calculate bending stresses, and SDOF equations were used to calculate hoop stresses, since Autopipe® does not calculate dynamic hoop stresses. The bending stresses exceeded the hoop stresses, but both hoop and bending stresses exceeded the fatigue strength of the ductile iron piping, which was not corroded. Numerous piping failures occurred; some near the end of piping, some near one near a pipe bend as shown below. Several other piping failures occurred in the middle of longer pipelines, where hoop stresses were the primary stresses in 20 foot long, bolted flange, piping sections. In all cases pipes were sheared perpendicular to the pipe wall

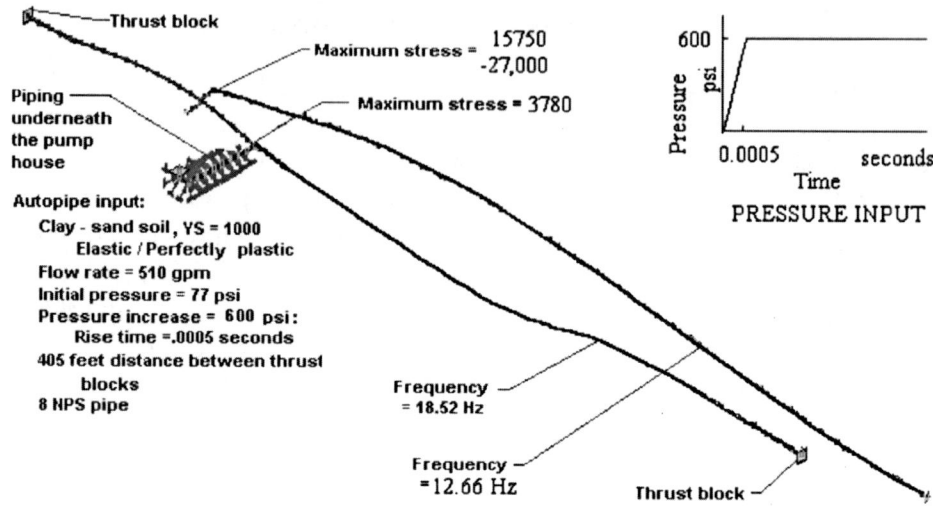

Figure 9.13 FEA model of bending stresses in ductile iron piping.

in a brittle (fatigue) manner, where only slight ductile deformation was observed near one side of the pipes. I inspected piping failures in the trenches that were dug to make repairs. Piping was repaired in place and not removed for further analysis.

7.2 Hoop stress piping failures in 2 inch diameter, carbon steel piping

Most of the failures in these cooling systems occurred on cooling coils inside the waste storage tanks, where the radioactive environment precluded direct inspection of piping failures. Consequently, an earlier study in the 1970's that concluded corrosion was the problem remained in effect for decades. As my research progressed, SRS implemented corrective actions in one facility, but elected to forego these actions in a second facility due to costs and the fact that the second facility was due for earlier decommissioning, which seemed reasonable. In this second facility, underground ductile iron piping was destroyed at a repair cost of approximately 5 million dollars. My theories were proven by practical experience, and I then developed water hammer classes and trained several hundred SRS operators and engineers how to properly operate systems to prevent water hammer damages.

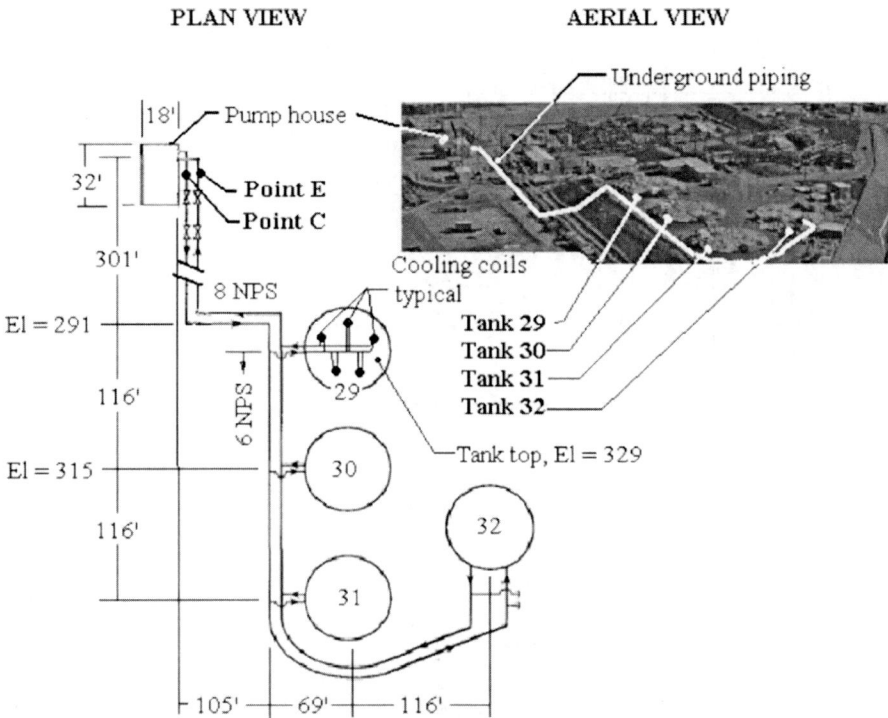

Figure 5.12 Partial cooling system layout.

Figure 5.14 Cooling coil damages.

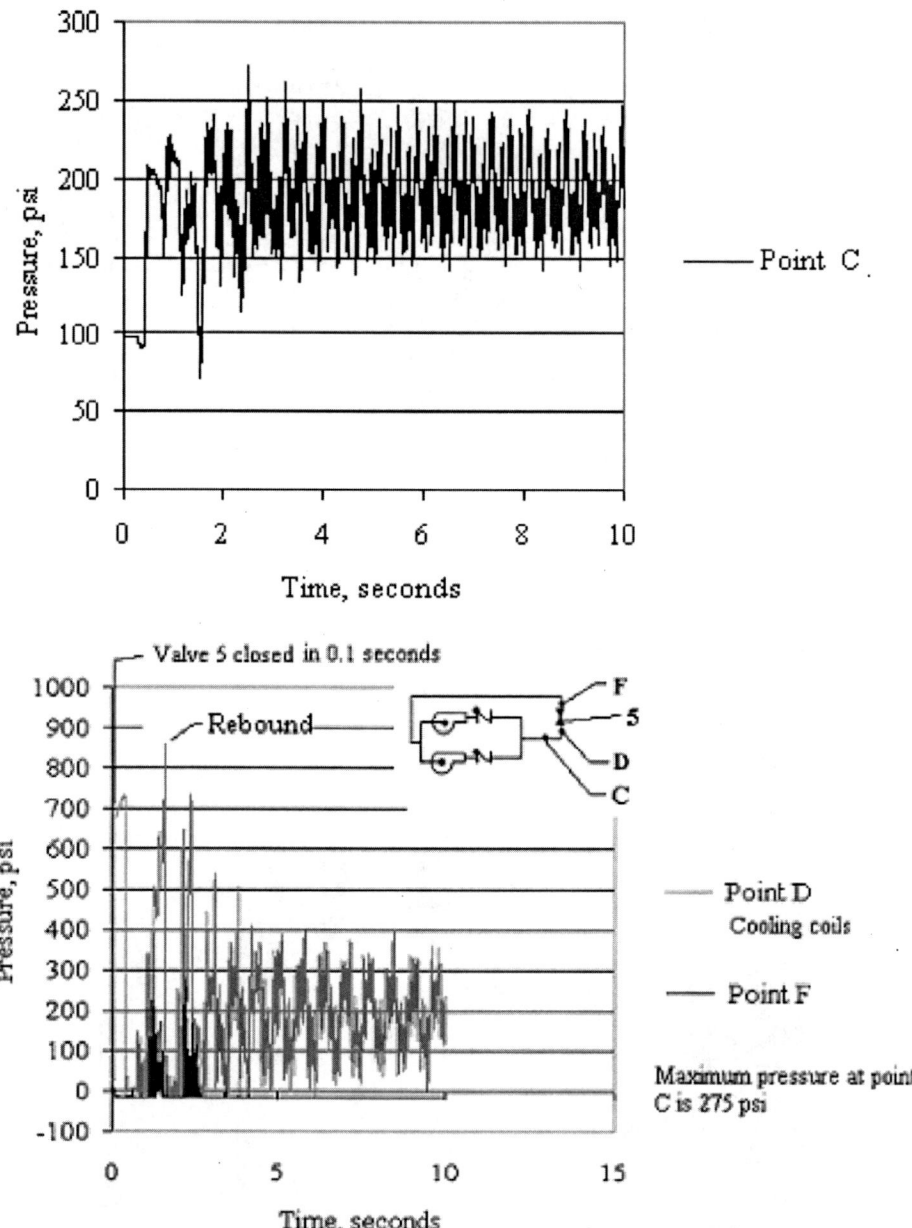

Figure 5.41 Reflected pressure wave effects due to single valve closure.

For this particular calculation, a DLF = 2 was assumed for hoop stresses and fatigue limits were shown to be exceeded by this analysis. Note that this approximation is less than the expected DLF < 4. There were also loads due to bending, where salt columns formed on the coils due to additional weight. The number and magnitude of water hammer loads was impossible to discern. Eliminating piping failures by eliminating hammers was sufficient proof of theory. The layout for part of the facility, a photo of piping leaks, and one of many water hammer calculations are provided in Figures 5.12, 5.14, and 5.41. Details are provided in my book.

8. Plastic deformations and pipeline explosions

Plastic deformations and explosions of pipelines are very significant, but have not been addressed to date in Appendix B.2. An example of plastic deformation is shown below, where 24 inch piping was knocked from its support due to water hammer (Figure 6.5). The steam blast from a 24 inch diameter, cast iron, exploding steam pipeline in New York City injured 45 people, and killed one person from a heart attack. A photo is shown below. There are many other examples of exploded and plastically deformed piping and vessels in the literature.

Figure 6.5 Pipe failure during initial system startup of steam system at SRS.

9. DLF's for plastic deformations

The transition from elastic to plastic strains in piping significantly affects the piping response (Figures 8.66 and 8.67). Theoretically, the DLF varies between 1 and 2, but significant damping during plasticity probably limits the DLF closer to a value of 1 (Figure 7.26). Some research indicates that FEA models describe the bounding stresses and strains in piping reasonably well when piping is subjected to internal pressure spikes (Figure 8.74 for example).

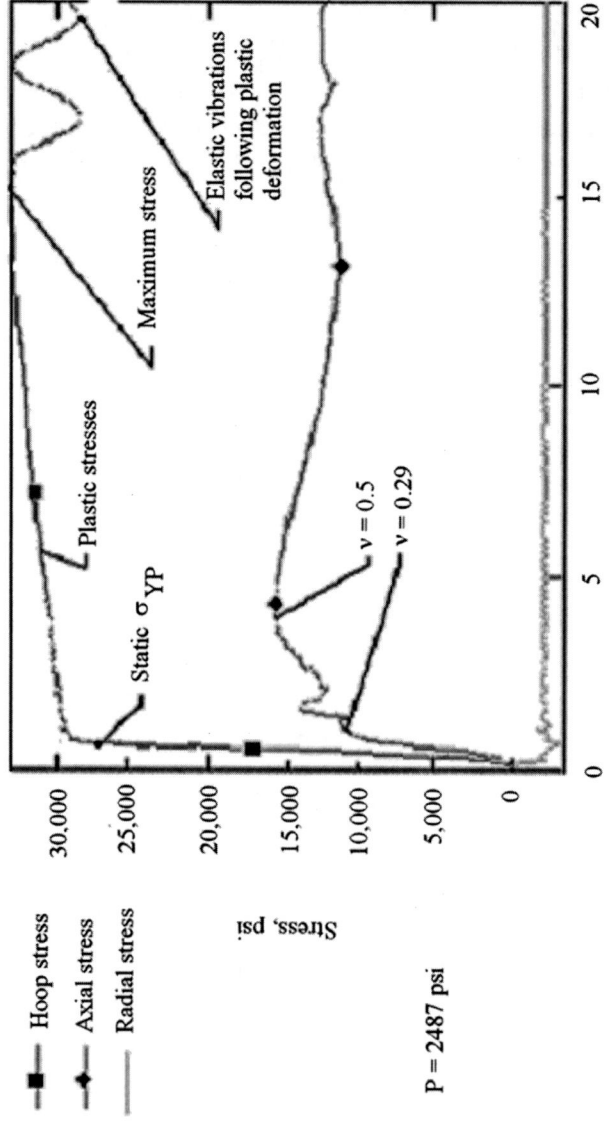

Figure 8.66 Plastic hoop stresses in 6 inch steel pipe without strain rate effects.

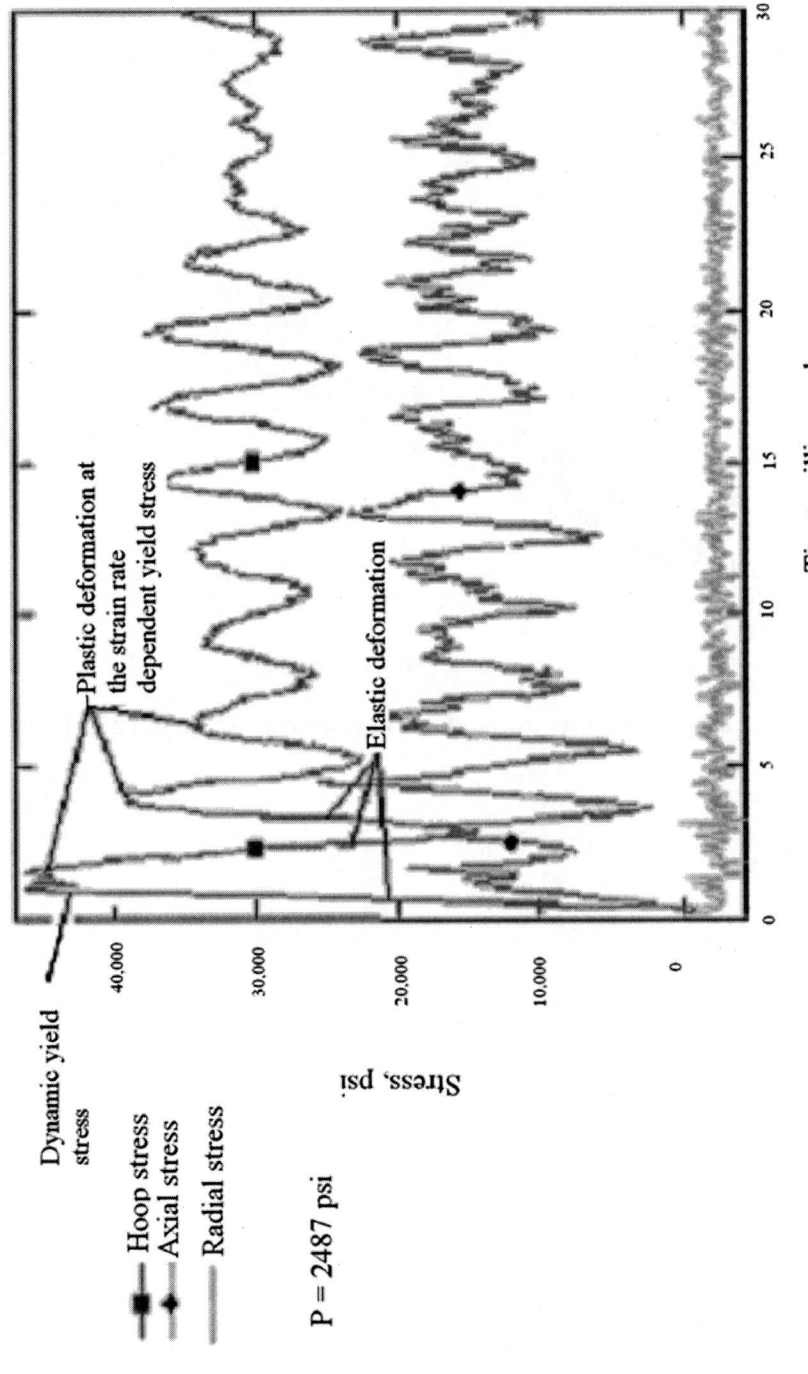

Figure 8.67 Plastic hoop stresses in 6 inch steel pipe with strain rate effects.

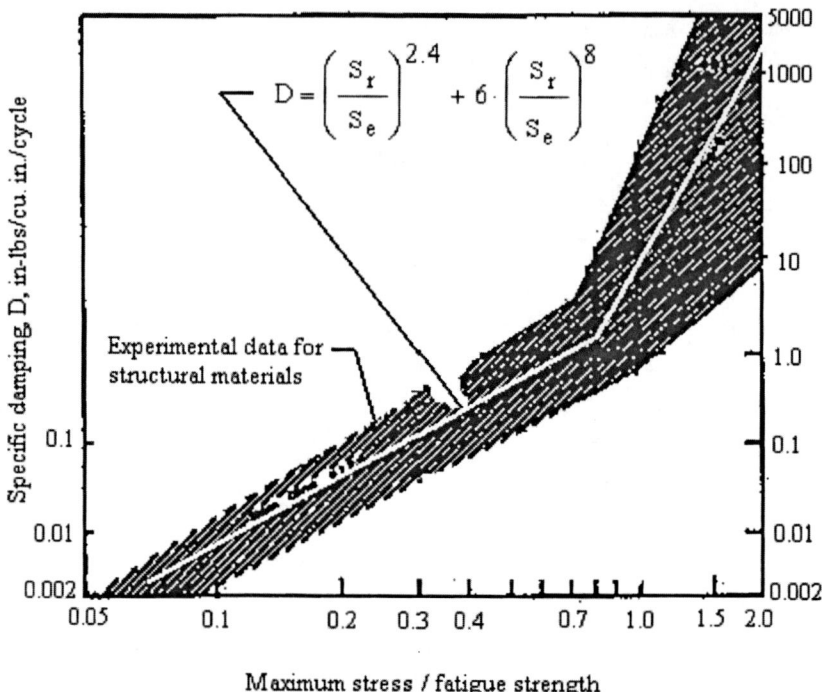

Figure 7.26 Damping for structural materials (Lazan [216]).

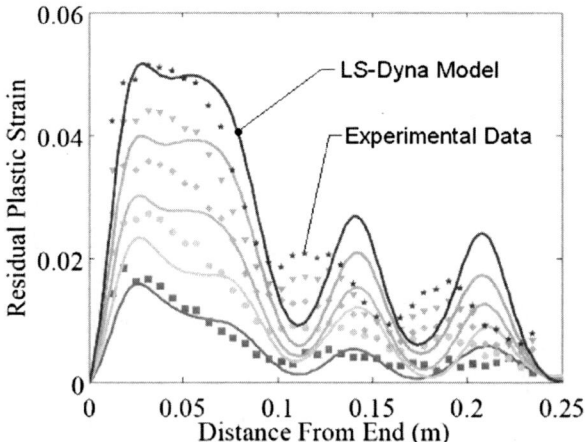

Figure 8.74 Residual plastic strain: LS-Dyna® FEA calculations compared to experiment for a linear strain hardening aluminum material, including strain rate effects on yield strength (Shepherd et al. [235]).

II.B. Appendix B.2:

Design of Piping Systems for Dynamic Loads from Fluid Transients

The following presentation provides a revision to an earlier version of a new piping standard, ASME B31D. The ASME B31 Mechanical Design Committee balloted that all water hammer equations, reflected wave equations, stress equations, and dynamic load equations, along with figures supporting those equations, would be removed from the Standard. This revised Standard would present a more simplified, qualitative discussion of fluid transients. Even so, this deleted revision to the new Standard provides substantial guidance to piping designers, and is therefore included as an Appendix to this book.

Foreword

Consideration of fluid transients in piping systems is sometimes required by Codes or by regulations, or it may be voluntarily considered for safe and reliable operation.

Of primary concern, damages to piping systems and pipelines due to fluid transients continue to occur. Design guidance and clear explanations and understanding are required for fluid transients and resultant structural responses of piping systems and pipelines.

Of secondary concern, fluid transients occur to some extent in all piping systems, where many transients have negligible effects on system operations and reliability. The need to identify acceptable fluid transients needs to be addressed, as well as identifying corrective actions for transients that may lead to damage and injuries.

While fluid transient loads are mentioned in the ASME B31 Pressure Piping Code, and allowable stresses are provided for occasional loads, there is a need to provide guidance for design of new piping systems, and there is a need to provide guidance for retrofit and operation of existing systems.

Introduction

The ASME B31 Code for Pressure Piping consists of a number of individually published Sections and Standards, each an American National Standard, under the direction of the ASME Committee B31, Code for Pressure Piping.

Rules for each Standard provide standardized guidance for a specific task found in one or more B31 Section publications, as follows:

(a) B31D, Design of Piping Systems for Dynamic Loads from Fluid Transients, in-process.

(b) B31E, Standard for the Seismic Design and Retrofit of Above-Ground Piping Systems, establishes a method for the seismic design of above-ground piping systems in the scope of the ASME B31 Code for Pressure Piping.

(c) B31G, Manual for Determining the Remaining Strength of Corroded Pipelines, provides a simplified procedure to determine the effect of wall loss due to corrosion or corrosion-like defects on pressure integrity in pipeline systems.

(d) B31J, Standard Test Method for Determining Stress Intensification Factors (i-Factors) for Metallic Piping Components, provides a standardized method to develop the stress intensification factors used in B31 piping analysis.

(e) B31Q, Pipeline Personnel Qualification, establishes the requirements for developing and implementing an effective Pipeline Personnel Qualification Program.

(f) B31T, Standard Toughness Requirements for Piping, provides requirements for evaluating the suitability of materials used in piping systems for piping that may be subject to brittle failure due to low-temperature service conditions.

As stated in the Scope, this Appendix provides guidance to help understand the causes, and to prevent or mitigate the effects, of fluid transients in piping and pipeline systems.

1. Purpose

This Appendix provides guidance to help understand the causes, and to prevent or mitigate the effects, of fluid transients in piping and pipeline systems.

1.1 Scope

This Appendix applies to piping systems or pipelines which are in the scope of the ASME B31Codes. This Appendix addresses fluid transient effects (also commonly referred to as water hammer). It does not address (a) flow-induced vibration effects such as those caused by steady pressure pulsing or vortex shedding and (b) pressures due to explosions within piping are outside the scope of this Appendix.

1.2 Terms and definitions

autofrettage, intended plastic straining of a predetermined portion of the inside pipe diameter during pipe fabrication to improve resistance to fatigue failure caused by pressure variations.

centrifugal pump, a pump that converts energy to velocity using a rotating impeller. A typical example is pump with rotating, radial vanes.

check valve, a valve which permits flow in only one direction.

design pressure, the pressure, in combination with the coincident temperature, that governs the wall thickness or pressure rating of the piping component. It is not the pressure capacity of the pipe. The maximum permitted normal operating pressure as defined in the applicable B31 Code.

dead head pressure, static pressure caused by a pump when a downstream valve is closed to prevent flow through the operating pump.

dynamic load factor, a factor used to multiply static loads so they may be used in a structural piping system analysis to approximate the maximum forces, strains and stresses that result due to the application of time varying loads.

endurance limit (fatigue limit), the maximum fully reversing stress that a material can theoretically withstand for an infinite number of cycles.

fluid transient, flow rate, density, and pressure changes in a piping system caused by changes in operating conditions.

head, specific energy at any point in a piping system typically measured in feet (or meters) of water.

hydraulic grade line, the height of fluid located at each point in a piping system.

method of characteristics, a computer simulation which uses a finite difference technique by dividing the pipe system into a number of reaches, and then solving equations of motion and continuity simultaneously.

modulus of elasticity, linear relationship between stress and strain in a mechanical body such as piping or valves, which is typically measured in a tensile test.

operating pressure, system operating pressure at the onset of a fluid transient.

pipe losses, See system resistance.

plastic deformation, permanent deformation due to an applied load when the yield strength is exceeded.

positive displacement pump, a pump that converts energy to pressure by compressing a liquid. A typical example is a piston pump.

pump shut-down, de-energizing the pump.

ratcheting, occurs when the shakedown limit is exceeded, and plastic deformation increases during each cycle as plastic deformations ratchet, or increase, through the pipe wall as cycling continues. Also, a stress-related condition which occurs in the presence of varying and non-varying loads where plastic strain increases during each applied cycle of the varying load.

relief valve, valve used to limit pressure in liquid filled systems.

run-out, free flow condition of a pump with a minimal discharge piping length.

safety valve, valve used to limit pressure in gas or vapor filled systems.

system resistance, frictional resistance to flow through valves, piping, and other components.

water hammer, a term used to describe fluid transients, when banging sounds are sometimes heard as check valves slam shut or piping strikes adjacent structures. The terms water hammer and fluid transient are frequently used interchangeably, but fluid transient is the generally accepted technical term, since it describes any pressure or flow change in a system.

wave speed, the speed of a wave inside fluid filled piping.

yield strength (yield stress), onset of plastic deformation in a material as defined by the applicable Code.

2. Materials

This Appendix applies to metallic piping systems. Strain rate effects on yield strengths are outside the scope of this Appendix, but higher strain rates increase the yield strength of materials.

3. Classes of fluid transients

(a) This Appendix is divided into the following general classes for simplicity and organization of presentation:
1. Liquid filled systems.
2. Liquid-non-condensable gas filled systems.
3. Liquid-condensable vapor filled systems.
4. Vapor, or non-condensable gas, filled systems.

These four classes of fluid transients are differentiated here to simplify the discussion, although a system may evolve from one class to another during operation. For example, a system may be designed to operate solely as a fluid filled system during normal operation but may occasionally operate with vapor or gas pockets in the system during filling, refilling, start-up, or shut-down. A steam system is intended to operate as a vapor system, but condensate accumulation during shut-down or operations may cause the system to act as a liquid-vapor when restarted or during other off-normal conditions, such as when a steam trap fails.

(b) Classification is further complicated by the fact that different types of flow conditions occur during fluid transients. Again for simplicity, these different conditions are defined here as slug flow transients, pressure wave transients, and condensate induced water hammer.
1. Pressure wave transients occur where pressure waves travel within the fluid contained in a pipe system. However, pressure wave transients may also be produced when a vapor pocket collapses in a liquid filled system, which is a two phase flow condition.
2. Slug flow is described as the motion of a large slug, or slugs, of fluid though a pipe system, where large piping deflections may occur. However, pressure waves may occur within slugs when slugs impact closed valves or piping ends.
3. Condensate induced water hammer occurs when waves are induced during the collapse of vapor in two phase flows. This type of fluid transient is basically a pressure wave transient within the fluid. When condensate induced water hammer occurs, the system is partially filled with a vapor or gas. The fluid need not be water for the process of condensate induced water hammer.

(c) This Appendix may be used to aid the designer during system design and piping layout, by indicating configurations and operations that tend to result in fluid transients. It may also be used to investigate problems in operating systems, by guiding the field engineers or operators in their investigations.

4. Fluid transients in liquid filled piping systems

(a) Systems completely filled with liquid are subject to transients when valves open and close or pumps start-up or shut-down. Historically, graphic solutions were available to describe pump and valve operations, but the method of characteristics is typically applied to computer generated models to describe pressure transients, assuming one-dimensional flow characteristics. In general, the equations presented here provide simplified estimates of the maximum pressures obtained due to different operating conditions. Simplified hand calculations may be performed in some cases, but computer simulations are recommended in many cases when evaluating fluid transients, and when computer simulations show the design to be inadequate or damaging to the system, redesign of the system or other corrective actions are recommended to mitigate occurrence of the transient. Also, additional testing may be performed as required.

(b) Transient loads on the piping and resultant deflections are caused by transient pressures at elbows, laterals, orifices, tees, etc. caused by the incident pressure waves. Transient loads are additive to coincident operating loads, and are affected by system dynamics (see para. 8.4).

(c) Increased stresses due to pressure surges may add or subtract to the initial stress present during steady state operations.

(d) This Appendix provides simplified screening criteria based on sudden actuations of valves, pumps, and regulators to determine if further analysis is required for a piping system. Many of the pressures presented by equations herein may be used to determine the maximum pressure magnitudes obtained by suddenly closing or opening valves, starting or stopping pumps, or suddenly opening steam valves or regulators into piping systems. An implicit assumption for the use of dynamic load factors is that the piping response may be represented by a single modal response when the piping is subjected to a single fluid transient. However, in many cases computer simulations are required to analyze transients, where spectrum analysis and fluid transient analyses need consideration. While sudden actuation of components is a reasonable assumption in some cases, sudden actuation is not always the case. In fact, increasing the actuation time is one technique to reduce pressure surges in systems.

(e) For the case of valve closures, closure times of $20 \cdot L/a$ are considered adequate for many systems to prevent piping damage by reducing the dynamic effects significantly enough that static design conditions may be assumed. Shorter times may be adequate in some cases, but further analysis would be required. Longer closure times may be required to prevent relief valve or safety valve actuation, depending on the system design.

(f) Screening criteria are also presented here with respect to piping stresses. Dynamic load factors are presented herein to describe the maximum stresses in piping due to suddenly applied fluid transients, which can be described by step pressure increases. Again, some cases are adequately represented by these simplifying assumptions, while many are not. The effects of damping and rates of loading are not considered in detail here, but these effects will reduce the dynamic load factor. Pressure transient other than step pressure increases may result in lower or higher pressures. As examples, linearly increasing pressures can lead to pressures lower than sudden step pressure increases. Harmonically applied (sin or cosine) applied pressures, or forcing functions, may lead to much higher pressures at resonant conditions, where the forcing function frequency approaches the piping frequencies.

(g) Overall, the simplifying assumptions provide screening criteria to determine the maximum loads on piping and supports caused by fluid transients caused by step increases in system pressure. Stresses or strains can then be evaluated by using the design rules of the applicable B31 Code or Standard. Additional analysis to analyze specific systems is outside the scope of this Appendix.

(h) Following the recommendations described in this Appendix will mitigate piping system damages. However, minimizing fluid transients will prevent the need for mitigating actions. In other words, if the system is not hammered, design requirements to overcome hammering will not be required.

4.1 Transients due to valve openings and closures

4.1.1 Pressure surges due to sudden valve openings and closures

(a) For a suddenly (instantly) closed valve, the maximum upstream pressure surge magnitude (ΔP), or head (Δh) with respect to the operating pressure (P_0), is described approximately by

$$\Delta h(ft) = \pm \frac{a(ft/sec) \cdot \Delta V(ft/sec)}{g_c(ft/sec^2)} \quad \text{[Refs. 1–3] [Eq. 1]}$$
$$= \pm a \cdot \Delta V / g$$

$$\Delta h(m) = \pm \frac{a(m/sec) \cdot \Delta V(m/sec)}{g_c(m/sec^2)} \quad \text{[Refs. 1–3] [Eq. 2]}$$
$$= \pm a \cdot \Delta V / g$$

$$\Delta P(lbf/ft^2) = \pm \frac{\rho \cdot a \cdot \Delta V}{g_c}$$
$$= \pm \frac{\rho(lbm/ft^3) \cdot a(ft/sec) \cdot \Delta V(ft/sec)}{g_c(ft \cdot lbm/lbf \cdot sec^2)}$$
$$\text{[Refs. 1–3] [Eq. 3]}$$

$$\Delta P(N/m^2) = \pm \rho \cdot a \cdot \Delta V$$
$$= \pm \rho(kg/m^3) \cdot a(m/sec) \cdot \Delta V(m/sec)$$
$$\cdot (N \cdot sec^2 / kg \cdot m)$$
$$\text{[Refs. 1–3] [Eq. 4]}$$

where the minimum pressure is limited by the vapor pressure of the liquid in the pipe under vacuum conditions. Refer to Figure 4.1A (Valve 1). The surge upstream of the closed valve travels at a sub-sonic wave velocity, a. Pressure surges on the downstream side of the closing valve may also be affected by flow separation and vapor collapse (see para. 6.1).

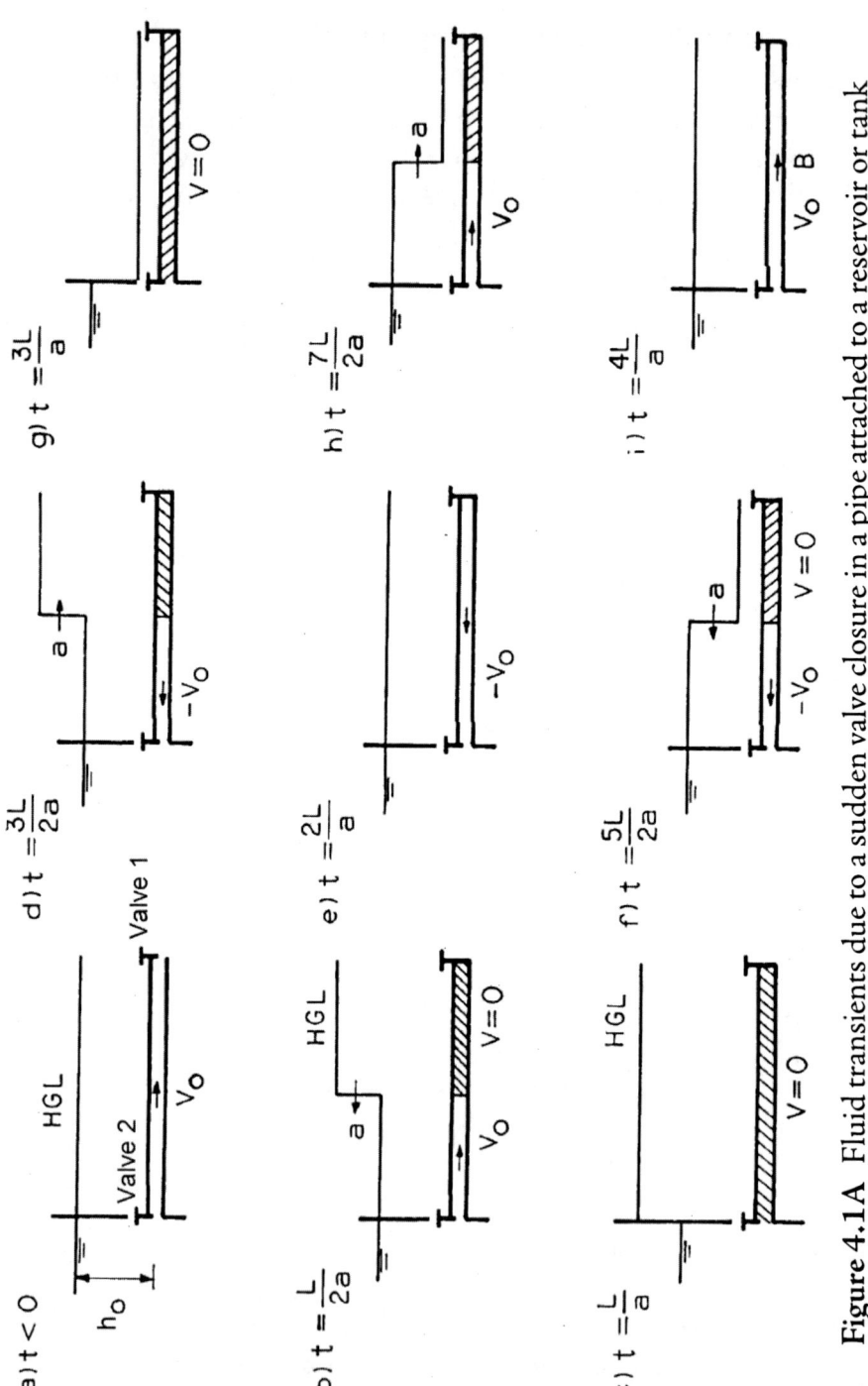

Figure 4.1A Fluid transients due to a sudden valve closure in a pipe attached to a reservoir or tank (pipe losses neglected) (Refs. 1–3, Courtesy of Sam Martin).

(b) The pressures, or heads, in a pipe following a valve closure are more complicated than indicated by Equations 1–4, as shown in the simplified system of Figure 4.1A. Waves travel back and forth in the pipe at a velocity, a, while the pressure surge behind the wave changes from positive to negative as the flow rate in the pipe stops and changes direction. The cycle repeats itself until system resistance stops fluid motion in the pipe (Figure 4.1B).

(c) Using Equations 3 or 4, the maximum pressure surge for Schedule 40 pipe is described by Figure 4.1C. It is emphasized that these pressures assume an instantaneous flow stoppage, which may be conservative but provide a bounding limit for pressure surge estimates.

(d) There is also a gradual pressure increase in the wake of the pressure surge, referred to as line pack. Figure 4.1B depicts line pack at a point in a piping system, as well as a comparison of experimental results to a method of characteristics calculation for the fluid transient.

(e) Pressures downstream of a closing valve in a pipeline may cause different maximum pressures than those observed upstream of a closing valve. For example, consider the system shown in

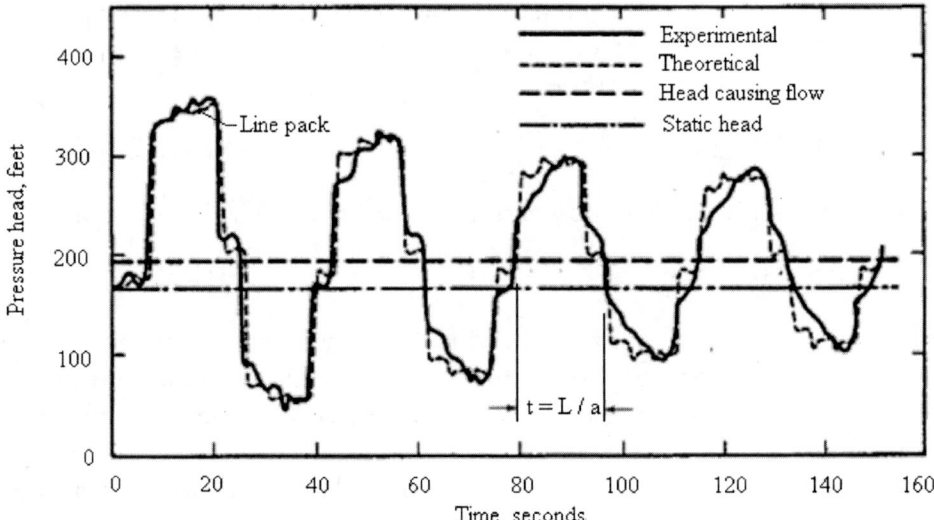

Figure 4.1B Comparison of theory to experiment (pipe losses considered) [Refs. 1, 3].

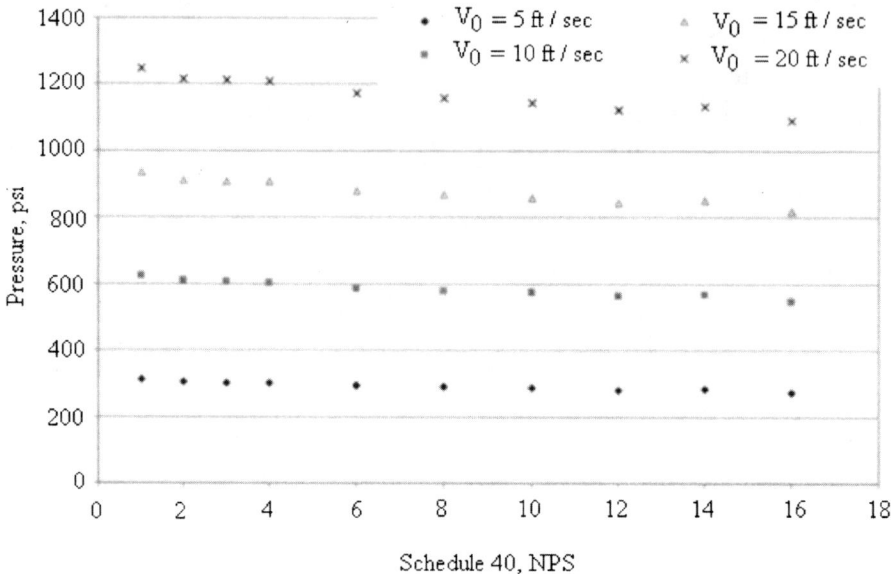

Figure 4.1C Maximum pressures due to sudden valve closures in schedule 40 pipe containing water (anchored, thick wall pipe approximation) [Ref. 3].

Figure 4.1A, but consider only Valve 2 while assuming that Valve 1 remains open. The maximum downstream head rise due to the instantaneous closure of Valve 2 equals

$$\Delta h = 0.22 \cdot h_0 \qquad \text{[Ref. 2] [Eq. 5]}$$

The maximum head drop downstream of the closing Valve 2 equals

$$\Delta h = 0.53 \cdot h_0 \qquad \text{[Ref. 2] [Eq. 6]}$$

In other words, the maximum head rise and drop upstream of a closing valve is described by Equations 1 and 2, while the maximum and minimum head downstream of a closing valve may be described by Equations 3 and 4.

Higher pressure surges may also be initiated on the downstream side of a closing valve due to flow separation and vapor collapse (see para. 6.1). Downstream pressures may also be reduced by system venting or draining.

(f) Referring to Figure 4.1A, the maximum head rise upstream of Valve 1 due to opening the valve equals

$$\Delta h = 0.23 \cdot h_0 \qquad \text{[Ref. 2] [Eq. 7]}$$

(g) Maximum pressure surge magnitudes initiated by valve operations in recirculating systems are also governed by Equations 3–7, but reflected waves throughout the system may further increase pressure surges.

(h) Planned valve closures are to be expected in operation, but off-normal conditions should also be anticipated. For example, unplanned transients may be caused by fail-to close (para 4.1.1a) or fail-to-open (paras. 4.1.3c, 5.2.1) valves on loss of electric power, or loss of air supply, depending on whether electric power or air pressure is used to actuate the valve.

(i) The definition of an instantaneous valve closure depends on specific systems. Equations 3 and 4 provide the approximate the maximum pressure surges due to valve closures, where the pressures may be reduced by valve closing characteristics. In particular, the magnitude of the pressure surge is affected by the closure rates of the valve, the valve design, pipe diameters, pipe lengths, number and types of fittings, and system friction. Equations presented in this Appendix may be used in some case, but in other cases computer simulations may be required to evaluate decreases in the maximum pressure caused by valve operations.

4.1.2 Pressure wave velocities in liquid filled piping

(a) Wave velocities, a, in a pipe are described by Equations 8–12, where the constant, c, describes limiting restraint conditions for piping. The actual value of the wave velocity in a pipe occurs between these two limiting values, and other values of c for different restraint conditions are available in the literature.

$$a = \sqrt{\dfrac{\dfrac{k(lbf/ft^2) \cdot g(ft/\sec^2)}{\gamma(lbf/ft^3)}}{1 + \left(\dfrac{k(lbf/ft^2)}{E(lbf/ft^2)}\right) \cdot \left(\dfrac{ID}{\overline{\overline{T}}}\right) \cdot c}} \qquad \text{[Refs. 1–3] [Eq. 8]}$$

$$a = \sqrt{\frac{\frac{k(N/m^2)}{\rho(kg/m^3)}}{1 + \left(\frac{k(N/m^2)}{E(N/m^2)}\right) \cdot \left(\frac{ID}{\overline{T}}\right) \cdot c}} \qquad \text{[Refs. 1–3] [Eq. 9]}$$

(b) Note that this velocity, *a*, is less than the speed of sound, or acoustic velocity, of the enclosed fluid, and that the acoustic velocity is affected by materials, pressures and temperatures, as shown in Figures 4.1D through 4.1F. Also, entrained air in the system can significantly change the wave speed.

(c) Figure 4.1E depicts wave speeds for thin wall pipes. For thin wall pipe anchored along the pipe length,

$$c = 1 - v^2 \qquad \text{[Refs. 1–3][Eq. 10]}$$

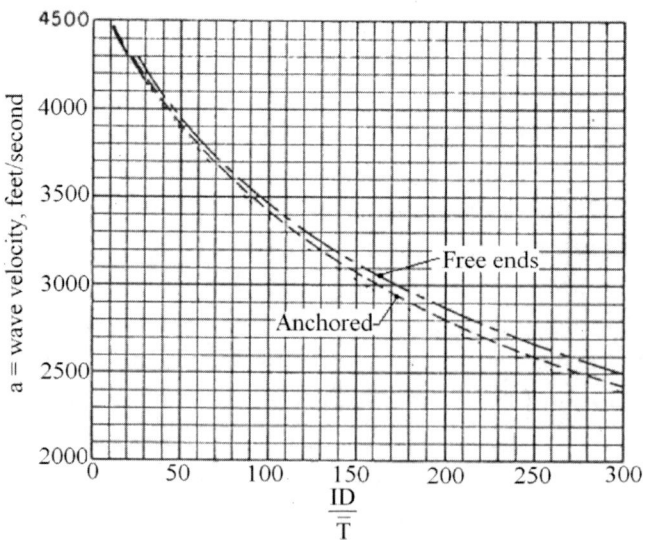

Figure 4.1D Wave velocities for water filled, thin wall piping at standard conditions (Ref. 2, reprinted with permission of Dover Publications).

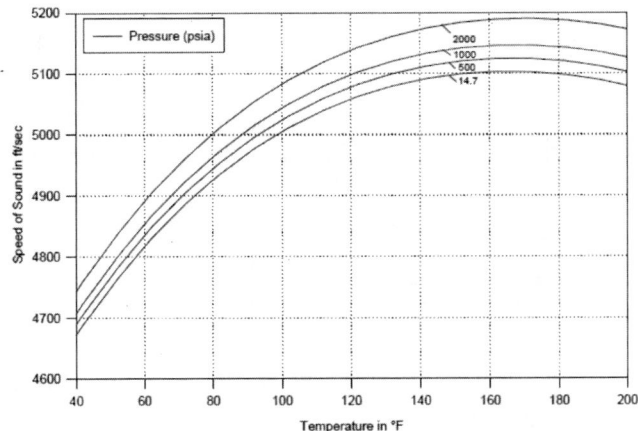

Figure 4.1E Speed of sound for water at different pressures and temperatures in thin wall piping (Ref. 4, Courtesy of Sam Martin).

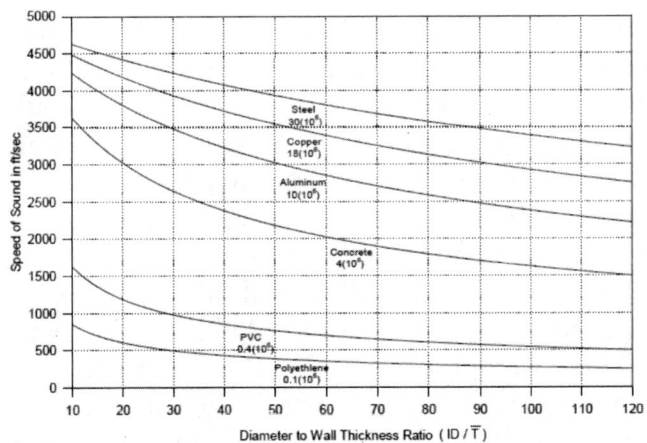

Figure 4.1F Wave velocities for different materials in a thin wall anchored pipeline (Ref. 4, Courtesy of Sam Martin).

The effects of piping material on wave speeds are shown in Figure 4.1F.
  (d) Thick wall approximations provide better accuracy for piping calculations. Simplifying assumptions are required to approximate wave speeds. For a thick wall pipe, assuming that one end is fixed and the other is free to move at the downstream end of the piping,

$$c = \frac{2 \cdot \overline{T}}{ID} \cdot (1+v) + \frac{ID}{ID + \overline{T}} \cdot \left(1 - \frac{v}{2}\right) \qquad \text{[Refs. 1–3] [Eq. 11]}$$

(e) For a thick wall pipe, and assuming that it is restrained along its entire length (anchored),

$$c = \frac{2 \cdot \overline{T}}{ID} \cdot (1+v) + \frac{ID}{ID + \overline{T}} \cdot (1 - v^2) \qquad \text{[Refs. 1–3] [Eq. 12]}$$

(f) Wave velocities in liquids are reduced when vapors or gasses are entrained in solution, due to the increase in compressibility, and are increased when solids are entrained in solution. In most cases, the entrapped air in solution has little effect, but the effects on wave speed where air is forced into the system should be evaluated.

4.1.3 Reflected and transmitted waves in liquid filled piping

(a) Waves initiated in piping systems are further complicated by reflected and transmitted waves, which occur at all transitions in diameter (reducers) or transitions in pipe material where densities or elastic moduli vary significantly at the transition (Figures 4.1G–4.1I) Additionally, fully reflected waves occur at closed ends in pipelines. Reflected waves may increase the pressure in the pipeline, and failures caused by reflected waves may occur close to the end of a pipe or near a transition in material or pipe diameter. The equations presented here provide simplified results, which do not include pressure reductions due to friction or energy losses due to pipe deformations, where pressures will be reduced due to friction similar to Figure 4.1B.

(b) A reflected wave at a closed end doubles the incident pressure wave magnitude at the closed end of a pipe. The magnitude of a reflected wave may be reduced if trapped air is present at the end of the pipe (see para. 5.1). Neglecting pipe losses, the maximum total pressure, P_{total}, near the end of pipe caused by an instantaneous valve closure in the system equals the sum

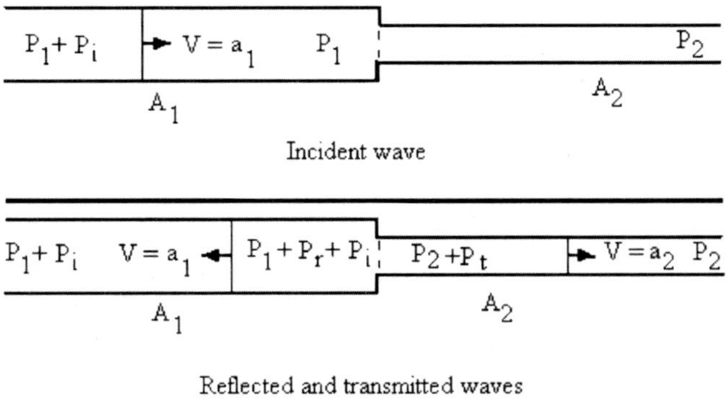

Figure 4.1G Pressure magnitudes due to reflections at reducers and changes in piping materials.

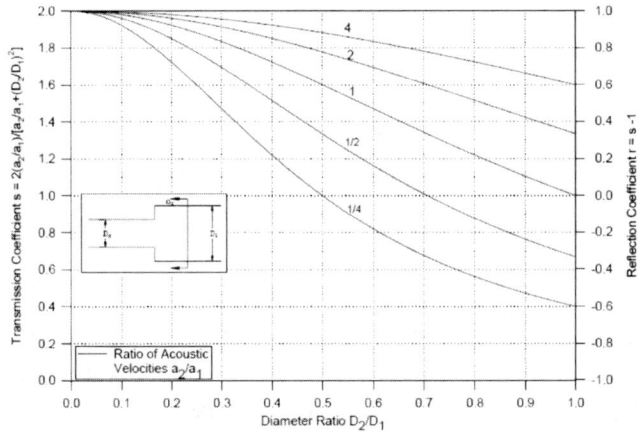

Figure 4.1H Wave reflections at a reducer when the diameter increases (Ref. 4, Courtesy of Sam Martin).

of two pressure wave magnitudes $(2 \cdot \Delta P)$ added to the initial pressure (P_1) in the piping, such that

$$P_{total} = P_1 + 2 \cdot \Delta P$$

$$= P_1 \left(lbf/ft^2 \right) + 2 \cdot \frac{\rho \left(lbm/ft^3 \right) \cdot a \left(ft/\sec \right) \cdot \Delta V \left(ft/\sec \right)}{g_c \left(ft \cdot lbm/lbf \cdot \sec^2 \right)}$$

[Refs. 1–3] [Eq. 13]

64 Appendix B.2

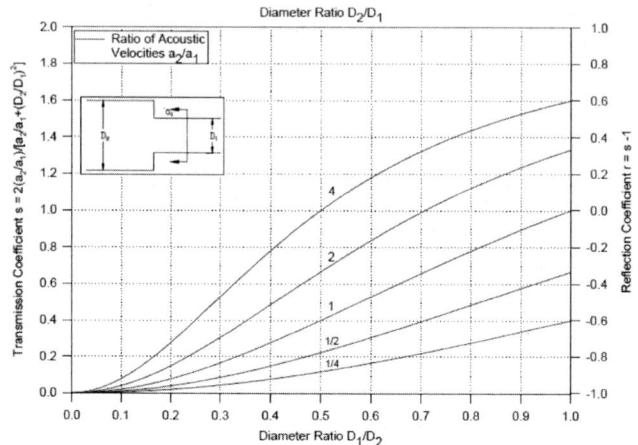

Figure 4.1I Wave reflections at a reducer when the diameter decreases (Ref. 4, Courtesy of Sam Martin).

$$P_{total} = P_1 + 2 \cdot \Delta P$$
$$= P_1(N/m^2) + 2 \cdot \rho(kg/m^3) \cdot a(m/\sec) \cdot \Delta V(m/\sec)$$

[Refs. 1–3] [Eq. 14]

(c) When a valve is opened into a closed end pipe filled with liquid, the change in pressure due to the valve's sudden opening is doubled in the pipe due to wave reflections at the pipe end, where the incident wave of pressure magnitude, ΔP, reflects to yield a pressure of $2 \cdot \Delta P$, which is added to the initial pressure in the piping (P_1), such that.

$$P_{total} = P_1 + 2 \cdot \Delta P \qquad \text{[Refs. 2, 3] [Eq. 15]}$$

The total pressure in the piping downstream of the opening valve then equals the initial pressure in the downstream piping plus $2 \cdot \Delta P$.

(d) At a pipe transition in diameter or material, the relationship between the pressure magnitudes of the incident (P_i), transmitted (P_t), and reflected pressure waves (P_r) may be expressed as

$$P_t = P_i \cdot \frac{\dfrac{2 \cdot A_1}{a_1}}{\dfrac{A_1}{a_1} + \dfrac{A_2}{a_2}} \qquad \text{[Refs. 2, 3] [Eq. 16]}$$

$$P_r = P_i \cdot \frac{\dfrac{A_1}{a_1} - \dfrac{A_2}{a_2}}{\dfrac{A_1}{a_1} + \dfrac{A_2}{a_2}} \qquad \text{[Refs. 2, 3] [Eq. 17]}$$

where A_1 and A_2 are the cross-sectional areas, and a_1 and a_2 are the wave speeds.

(e) The pressure magnitudes of the incident, transmitted, and reflected waves at a tee, or branch intersection, may be expressed as

$$P_t = P_i \cdot \frac{\dfrac{2 \cdot A_1}{a_1}}{\dfrac{A_1}{a_1} + \dfrac{A_2}{a_2} + \dfrac{A_3}{a_3}} \qquad \text{[Refs. 2, 3] [Eq. 18]}$$

$$P_r = P_i \cdot \frac{\dfrac{A_1}{a_1} - \dfrac{A_2}{a_2} - \dfrac{A_3}{a_3}}{\dfrac{A_1}{a_1} + \dfrac{A_2}{a_2} + \dfrac{A_3}{a_3}} \qquad \text{[Refs. 2, 3] [Eq. 19]}$$

where a_3 and A_3 are wave speeds and areas.

(f) Reflected and transmitted waves occur at all transitions and tees in a system (Figure 4.1J). Both positive and negative pressures may be reflected or transmitted at piping and tank intersections with the piping. Computer simulations are frequently required to fully understand complicated system performance, where the method of characteristics is an accepted simulation technique. Even so, Equations 18 and 19 provide simplified calculations for many designs.

66 APPENDIX B.2

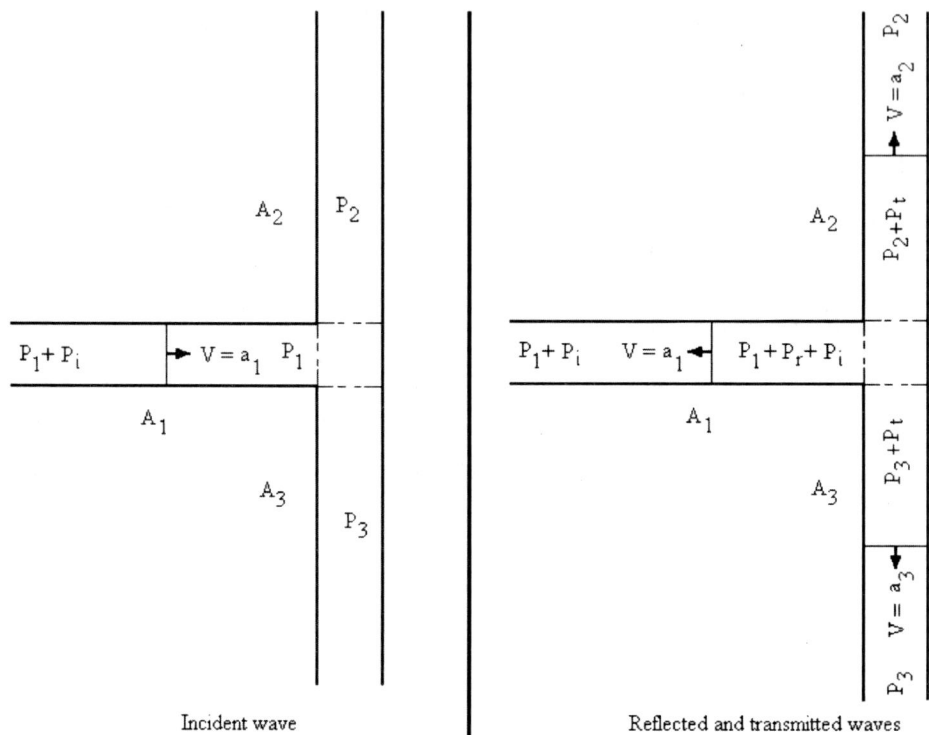

Figure 4.1J Pressure magnitudes due to reflections at pipe intersections.

4.2 Transients during centrifugal pump start-up and shut-down

4.2.1 Centrifugal pump operations

(a) When a single centrifugal pump starts or shuts down, the maximum pressure surge at the pump is limited by, but is typically much less than, Equations 3 and 4. In fact, during pump start-up, pump transient pressures are frequently less than twice the operating pressure of the pump. Changes in flow rate and pressure are transmitted throughout the system at the wave velocity, a. hand calculations and graphic techniques typically not describe the system adequately. The method of characteristics is the recommended method to compute the transient, since pressure surges are complicated by numerous factors during pump transients, such as system resistance, piping configuration, pump impeller inertia, and motor inertia.

(b) These effects are highlighted by the following examples to be considered when evaluating Centrifugal pump systems.

1. When pumping uphill between two reservoirs or tanks, flow slows down and reverses direction through a pump during pump shut down, and the pump then speeds up to the initial operating speed in the reverse direction to act as an electric generator. Check valves are installed at pump discharges to prevent flow reversal. If the pipeline is long enough, the pump may act as closing valve as described by Equations 3 and 4.
2. In a long pipeline, a pump may act the same as an opening or closing valve when pumping uphill.
3. When pumping in a recirculating system, the liquid usually coasts to a stop following pump shut-down, when the pump has negligible inertia with respect to the liquid momentum.
4. When a pump is started against a long, closed end, liquid filled pipe or a pipe with significant pipe losses, reflected waves may double the pump dead head pressure at the end of the pipe. For this case, the pump acts like a valve opening into a closed end, liquid filled pipe.
5. When pumping downhill between two reservoirs, flow separation may occur at the pump discharge and should be evaluated.
6. When pumping downhill to an empty tank or atmosphere, the pipe may drain and fill with air following pump shut-down. The method of characteristics cannot model this condition.
7. When one of two or more parallel pumps is shut down, a fluid transient may be caused by the slamming of the check valve on the slowing pump. The maximum pressure is described by Equations 3 and 4, but pump flow rates may be significantly higher since the pump operates at run-out conditions. The pressure surge may also be mitigated by check valve characteristics.

4.2.2 Operations of centrifugal pumps in parallel

(a) When centrifugal pumps are operated in parallel, transients occur when one pump is stopped while the others continue to operate. Figure 4.2A provides a schematic to consider a two pump system. If one pump is stopped while the other continues to run, the operating pump will effectively induce flow at run-out conditions backward through the stopping pump until the check valve protecting the shut-down pump closes. This phenomenon, called check valve slamming, induces a wave into the system. The magnitude of the pressure surge depends on the inertia of the slowing pump, check valve performance, and the flow rate obtained as valve slamming occurs. Accordingly, computer simulations are recommended.

4.3 Operations of positive displacement pumps

Fluid transients in systems using positive displacement pumps may be complicated by pulsating flows induced by some pump types, which may cause unstable, and sometimes large, piping system vibrations. Stronger or thicker piping or additional pipe supports may be used, but the preference is to reduce the pulsation. To minimize pulsations, common corrective actions include suction flow stabilizers; surge suppressors, which are specially designed devices used to reduce liquid pulsations.

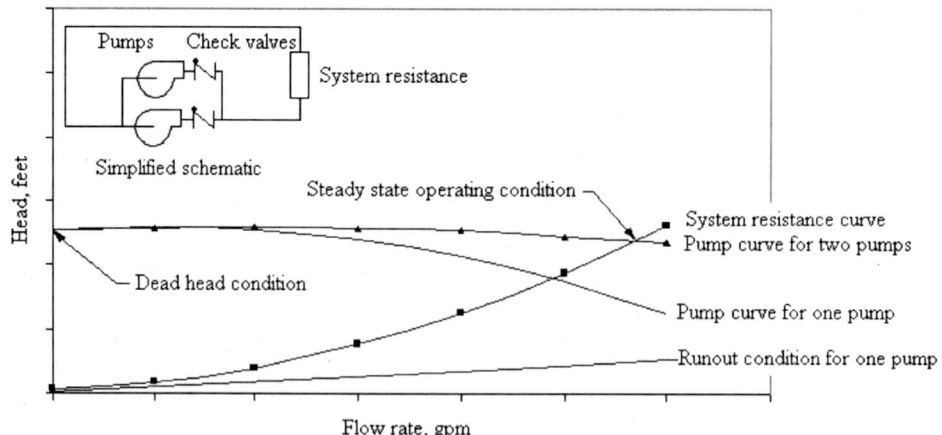

Figure 4.2A Steady state, parallel pump operations [Ref. 3].

5. Fluid transients in liquid-gas filled piping systems

Trapped gas in a liquid filled piping system can have either a beneficial or detrimental effect with respect to fluid transients, depending on the amount of trapped gas, the system design, and the location(s) of the trapped gas. Small quantities of gas can reduce transient effects. Large quantities of gas can cause significant transient effects.

5.1 Reduction in transient effects due to trapped air pockets

The use of a small quantity of a gas, such as air, in surge tanks and accumulators has long been a corrective action used to dampen fluid transients. Small quantities of trapped gas in piping systems will also reduce pressure surges caused by transients.

5.2 A comparison between sudden valve closure, slow closure, and air pocket effects

(a) The simplified results of a method of characteristics calculation are shown in Figures 5.2A, 5.2B, and 5.2C, for a recirculating system. The underground pipe is constructed of approximately 1000 feet of 6 and 8 inch Schedule 40, ductile iron pipe.

(b) In Figure 5.2A, pressure surges due to a sudden valve closure are shown.

(c) In Figure 5.2B, the effects of adding only one cubic foot of trapped air at valve are shown to reduce the reflected pressure surge from 850 psi to 275 psi.

(d) Figure 5.2C is added to show the comparative effects of a slow closing valve on the same system, where a 10 second ball valve closure time reduces the transient pressure to 92 psi.

5.3 Transient effects due to large air quantities in liquid filled piping systems

Filling a piping system that is initially filled with air is a common occurrence to consider the effects of large quantities of gas in liquid systems. Filling may occur during the first operation and subsequent refilling, or filling may occur every time the piping is used such as initially empty fire suppression systems.

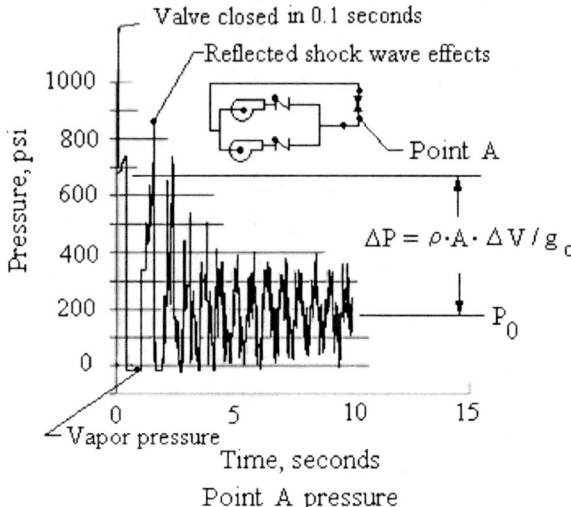

Figure 5.2A Fluid transient due to a sudden valve closure [Ref. 3].

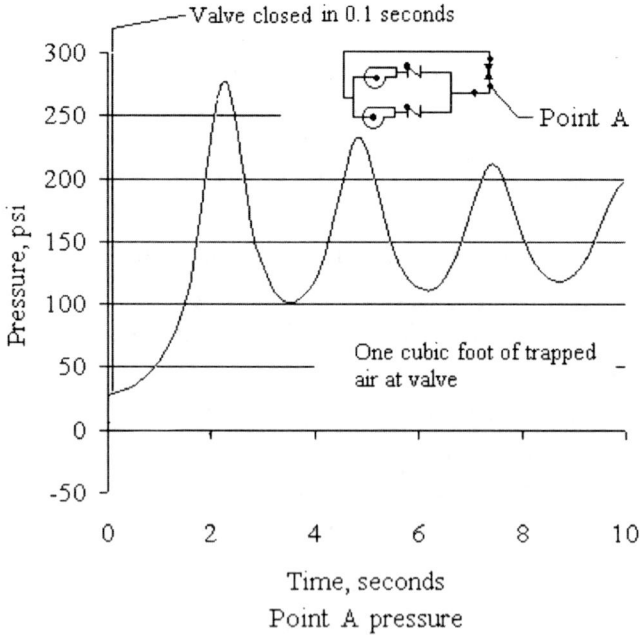

Figure 5.2B Air pocket effects on a fluid transient [Ref. 3].

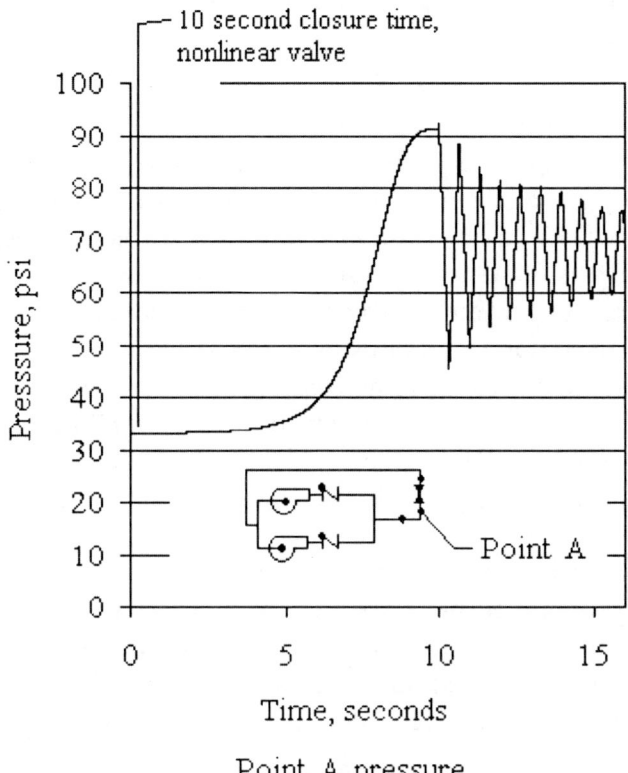

Figure 5.2C Effects of a slow closing valve [Ref. 3].

5.3.1 Transient effects due to liquid filling of air filled piping systems

If the discharge valve of an operating pump is suddenly opened into an empty pipe, the resistance to flow may be much less than design. Instead of operating at the design flow rate, the pump may operate near run-out (Figure 4.2A). The sudden increase in flow rate acts as a slug flow impinging on elbows or tees near the pump discharge. Loads are due to a change in momentum of the slug, rather than a wave transmitted in the system.

6. Fluid transients in liquid-vapor filled piping systems

6.1 Pressure surges due to void formation and vapor collapse

(a) Two types of liquid-vapor systems are presented here. The first type concerns a liquid filled system, which induces vapor formation during pump or valve operations. The second type concerns condensate induced water hammer (also referred to as steam hammer), which occurs for some conditions when condensation occurs in steam system piping.

(b) Loads, deflections, and stresses are caused by increased pressures induced by the impacts of liquid slug flows on collapsing vapor spaces. The vapor condenses when impacted by a liquid slug, and offers negligible resistance to the liquid flow rate during vapor collapse.

(c) The method of characteristics has been successful in predicting pressure surges for some, but not all, types of liquid-vapor transients. The discrete vapor cavity model is used along with the method of characteristics to describe liquid-vapor transients. The discrete vapor cavity model considers the vapor cavity to be a single cavity of vapor, and a comparison of the model to experiment is shown in Figure 6.1A.

(d) In general, the damaging effects of vapor collapse can be avoided by continuous operation of a system where practical, and when not practical steps should be taken to slowly shut-down or restart systems which contain vapor pockets. Steam traps and insulation should be maintained in working order.

6.2 Fluid transients due to vapor collapse in liquid filled piping systems

Transients occur when vapor pockets collapse in liquid filled systems. In some cases, vapor pockets, or voids, are formed downstream of closing valves, downstream of pumps during shut-down, and at high points in piping systems during transients. Transients may occur immediately following a valve closure or pump shut-down, or may occur at a later time when the valve is reopened or the pump is restarted.

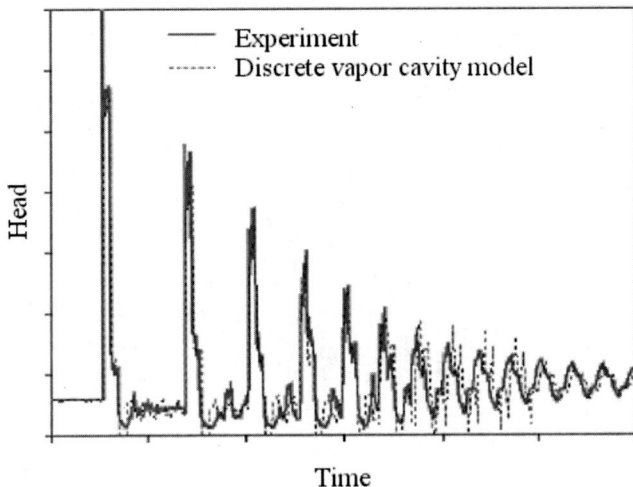

Figure 6.1A Comparison of experiment to the discrete vapor cavity model (Adamkowski and Lewandowski [193]).

6.2.1 Void formation and vapor collapse

(a) When the pressure in a pipe reduces to the vapor pressure of the liquid contained in the pipe, the liquid vaporizes. During void formation, this low pressure occurs when the liquid column in the pipe separates due to motion. When the column rejoins, the vapor collapses, and

(b) Waves are induced throughout the liquid filled system. Again, computer simulations are recommended, but when computer simulations are unavailable systems should be designed to prevent occurrence of vapor collapse.

(c) For example, if the pipe elevation in a system exceeds the hydraulic grade line, the liquid in the pipe can vaporize during a transient. Figure 6.2A can be used to describe a transient at a high point in a piping system immediately after a pump shutdown or valve closure, where h_b is the barometric head, h_v is the vapor pressure of the liquid in the pipe, and Q_1 and Q_2 are volumetric flow rates. Referring to Figure 5.2A, the difference increases between flow rates of Q_1 and Q_2 when flow is stopped at the valve. This difference in flow rates may cause vaporization at the high point at point 3, and may also cause vaporization at point 1 near the valve.

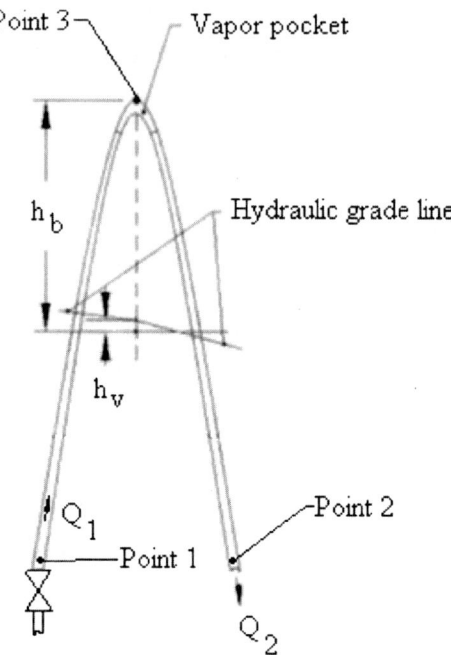

Figure 6.2A Column separation, void formation, and vapor collapse [Refs. 2, 3].

(d) If the piping end at point 2 is submerged in a liquid, vapor pockets will quickly collapse, as a liquid slug returns to close the voids. A wave is then formed, which has a maximum theoretical pressure equal to one half the magnitude of Equations 3 or 4, and is expressed as

$$\Delta P \left(lbf/ft^2 \right) = \frac{\rho \left(lbm/ft^3 \right) \cdot a \left(ft/sec \right) \cdot \Delta V \left(ft/sec \right)}{2 \cdot g_c \left(ft \cdot lbm/lbf \cdot sec^2 \right)} \quad \text{[Ref. 3] [Eq. 20]}$$

$$\Delta P \left(N/m^2 \right) = \frac{\rho \left(kg/m^3 \right) \cdot a \left(m/sec \right) \cdot \Delta V \left(m/sec \right) \cdot \left(N \cdot sec^2/kg \cdot m \right)}{2} \quad \text{[Ref. 3] [Eq. 21]}$$

(e) In general, computer simulations are recommended, using the method of characteristics, but velocities can sometimes be calculated using system curves along with pump curves, where detailed calculations are outside the scope of this Appendix. When computer simulations are unavailable, systems should be designed to prevent occurrence of vapor collapse.

(f) There are also cases where trapped vapor pockets remain at the top of a pipe, or below a valve near the top of a pipe, following an upstream valve closure or pump shut-down.
1. These pockets collapse when the system is re-pressurized by an upstream valve opening or pump restart, as shown in Figure 6.2B. The maximum pressures are governed by Equations 3 or 4. Computer simulations are frequently inadequate for this type of transient.
2. For piping systems where vapor accumulates at high points in the system after pump shut-downs systems, Equation 20 or 21 governs the pressure surge on re-pressurization.

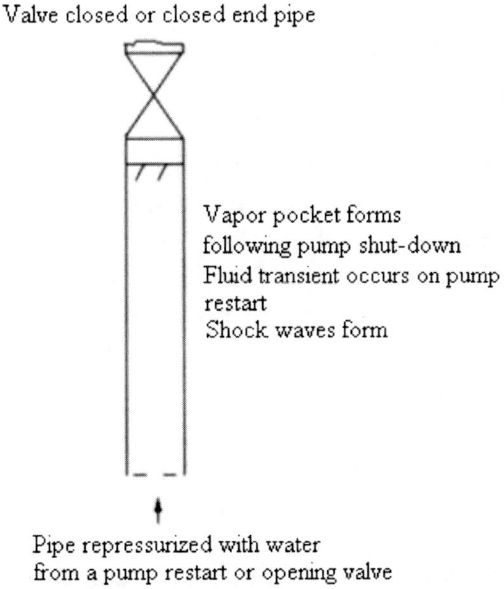

Valve closed or closed end pipe

Vapor pocket forms following pump shut-down
Fluid transient occurs on pump restart
Shock waves form

Pipe repressurized with water from a pump restart or opening valve

Figure 6.2B Filling a vertical pipe containing a trapped vapor space (Refs. 3, 5, reprinted by permission of EPRI).

Computer simulations are frequently inadequate for this type of transient.
3. Comparable conditions occur in steam condensate systems, where the primary difference is the lower vapor pressure which will more readily induce vapor pockets.

6.2.2 Damage mechanisms and corrective actions for vapor collapse in liquid filled systems

Typically, piping damage mechanisms occur within the liquid volume where pressure waves occur, and piping damage mechanisms include (but are not limited to):
1. Fracture and shear of brittle components.
2. Plastic deformation of ductile piping.
3. Valve and component leaks.

6.3 Condensate induced water hammer

There are numerous conditions where condensate can form in steam piping systems. Once formed, condensate and steam vapor may combine to create fluid transients. Descriptions for some of these transients follow. During this type of water hammer, multiple pressure waves may occur if multiple vapor pockets collapse.

6.3.1 Condensate induced water hammer in horizontal pipes

(a) Condensate induced water hammer occurs in a horizontal pipe when liquid accumulates below a vapor flow in a pipe. The water may initially be moving or stationary in the pipe. One cause of condensate accumulation occurs during steam system shut-down, where the closed system cools to form vapor and liquid. Another source of condensate is considered in Figure 6.3A. Both causes result in condensate induced water hammer.

(b) For Figure 6.3A:
1. Condensate slowly fills a pipe supplied with steam.
2. Unstable wave growth occurs.
3. Steam voids and slugs are formed.
4. Slugs accelerate.
5. Steam voids collapse due to slug impacts.
6. Waves and/or liquid pulses are formed.

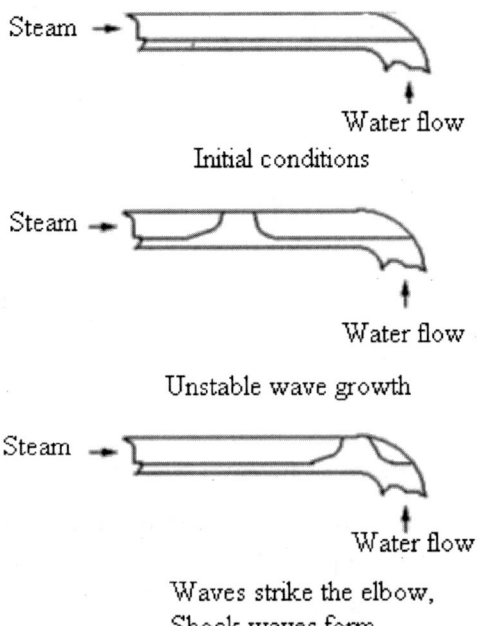

Figure 6.3A Condensate induced water hammer in a horizontal pipe (Refs. 3, 5, reprinted by permission of EPRI).

(c) Slugs of water may be propelled into the piping system when valves are suddenly opened to permit steam into piping where condensate has accumulated. Steam flashing may also occur, which can result in additional forces caused by water slugs. Computer simulations may be available, but when computer simulations are unavailable systems should be designed to prevent occurrence of condensate induced water hammer.

6.3.2 Water cannon
 (a) The water cannon mechanism is described using Figure 6.3B.
  1. Steam enters a cooler fluid.
  2. A valve is closed to throttle or stop the flow.
  3. Steam condenses as water rushes into the pipe.
  4. Vapor and liquid mix to form vapor pockets, which then collapse and cause pressure waves.
 (b) Computer simulations are available, but when computer simulations are unavailable, systems should be designed to prevent occurrence of water cannon.

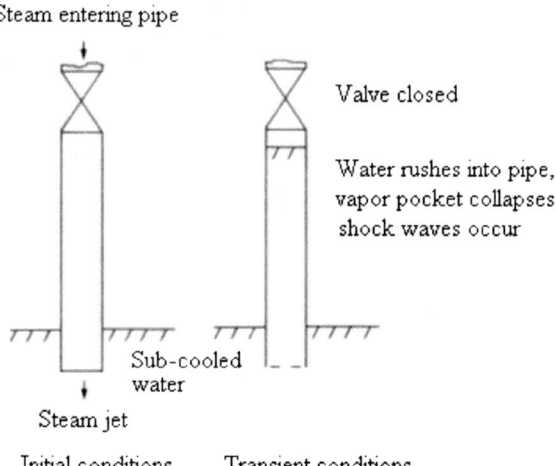

Figure 6.3B Water cannon, sub-cooled water with condensing steam in a vertical pipe (Refs. 3, 5, reprinted by permission of EPRI).

6.3.3 Damages and corrective actions for condensate induced water hammer
  (a) Condensate induced water hammer should be avoided. Observed damages due to condensate induced water hammer include (but are not limited to):
    1. Fracture and/or catastrophic failure of brittle components.
    2. Condensate induced water hammer resulted in an operator fatality due to failure of a brittle valve. In this case, an operator was subjected to steam vapors in an enclosed space. The steam displaced the air following explosion of a valve, and the escaping steam burned the operator's lungs. A pressure surge up to 3000 psig was estimated due to slug flow in this accident, where the system operating pressure was 110 psig.
    3. An explosion of a cast iron pipe also occurred due to condensate induced water hammer, and an eight foot diameter hole was blown up through a paved walkway.
  (b) Accidents due to previous plastic deformation in ductile piping systems.
    1. Piping and pipe supports have been bent and distorted.

2. Piping (24 inch diameter) has been knocked from pipe supports.
3. Shearing of flange bolts due to condensate induced water hammer resulted in fatalities due to personnel steam exposure.
4. Valve and component leaks have been attributed to condensate induced water hammer.
5. Other ductile piping failures may also be attributed to water hammer, but further investigation is required.

6.4 Slug flow

(a) Slug flow occurs in various systems, where a slug, or slugs, of liquid move through the piping. Forces may be exerted on elbows and fittings due to changes in fluid momentum, where

$$\Delta P \left(lbf/ft^2 \right) = \frac{\rho \left(lbm/ft^3 \right) \cdot V^2 \left(ft^2/sec^2 \right)}{2 \cdot g_c \left(ft \cdot lbm/lbf \cdot sec^2 \right)} \quad \text{[Eq. 22]}$$

$$\Delta P \left(N/m^2 \right) = \frac{\rho \left(kg/m^3 \right) \cdot V^2 \left(m^2/sec^2 \right) \cdot \left(N \cdot s^2/kg \cdot m \right)}{2} \quad \text{[Eq. 23]}$$

(b) Higher forces may also be observed when slugs impact the piping ends or closed valves, where the maximum pressure due to a slug flow impact at the end of a pipe or closed valve in a vapor filled system may be approximated using the velocity, V_s, of a slug during vapor collapse

$$V_s \left(ft/sec \right) = \sqrt{\frac{2 \cdot L_v \left(ft \right) \cdot P \left(lbf/ft^2 \right) \cdot g_c \left(ft \cdot lbm/lbf \cdot sec^2 \right)}{\rho \left(lbm/ft^3 \right) \cdot L_s \left(ft \right)}}$$

[Ref. 3] [Eq. 24]

$$V_s \left(m/sec \right) = \sqrt{\frac{2 \cdot L_v \left(m \right) \cdot P \left(N/m^2 \right)}{\rho \left(kg/m^3 \right) \cdot L_s \left(m \right) \cdot \left(N \cdot sec^2/kg \cdot m \right)}} \quad \text{[Ref. 3] [Eq. 25]}$$

where L_v is the length of the vapor space, L_s is the length of the slug, and P is the pressure applied to move the slug. The equivalent pressure magnitude of the pressure waves induced within the liquid slug following impact at a closed end pipe then equals

$$\Delta P\left(lbf/ft^2\right) = \frac{\rho\left(lbm/ft^3\right) \cdot a\left(ft/sec\right) \cdot V_s\left(ft/sec\right)}{g_c\left(ft \cdot lbm/lbf \cdot sec^2\right)} \quad \text{[Ref. 3] [Eq. 26]}$$

$$\Delta P\left(N/m^2\right) = \rho\left(kg/m^3\right) \cdot a\left(m/sec\right) \cdot V_s\left(m/sec\right) \cdot \left(N \cdot sec^2/kg \cdot m\right)$$
[Ref. 3] [Eq. 27]

Note also that dynamic effects may double the effects of the applied pressures determined by Equations 26 and 27, when a DLF is considered (See paragraph 8.2).

(c) Computer simulations are recommended since the transient pressures are more complicated than indicated by Equations 26 and 27. When computer simulations are unavailable, systems should be designed to prevent occurrence.

7. Fluid transients in vapor or gas filled piping systems

A discussion of transients in vapor and gas filled systems is limited here to those transients caused by valve openings. For a suddenly opened, or closed, valve, the applied pressure to the system (ΔP) equals the difference in initial pressures on either side of the actuated valve. The upstream pressure is limited to the critical pressure at sonic velocity. By way of example, a special case is related to safety valve operation (adapted from B31.1, Appendix II).

7.1 Pressure loads on safety valve installations

Pressure loads acting on safety valve installations are important from two main considerations. The first consideration is that safety valve installation causes membrane stresses in the pipe wall. The second consideration is that the pressure effects associated with discharge can cause loads acting on the system which create bending moments throughout the piping system. All parts of the safety valve installation must be designed to withstand the design pressures without exceeding the Code allowable stresses, and dynamic effects are considered during design. Only fluid transient loads are considered in this Appendix, and other loads are considered per the applicable standard.

7.2 Reaction forces on open discharge pipes

7.2.1 Forces on discharge elbows

(a) The reaction force, F_1, due to steady state flow following the opening of a safety valve includes both momentum and pressure effects. The reaction force applied is shown in Figure 7.2A, and may be computed by the following equation:

$$F_1(lbf) = \frac{W(lbm/sec) \cdot V_1(ft/sec)}{g_c(ft \cdot lbm/lbf \cdot sec^2)} + \left(P_1(lbf/ft^2) - P_a(lbf/ft^2)\right) \cdot A_1(ft^2) \quad \text{[Ref. 3] [Eq. 28]}$$

$$F_1(N) = W(kg/sec^2) \cdot V_1(m/sec) \cdot \left(N \cdot sec^2/kg \cdot m\right) + \left(P_1(N/m^2) - P_a(N/m^2)\right) \cdot A_1(m^2) \quad \text{[Ref. 3] [Eq. 29]}$$

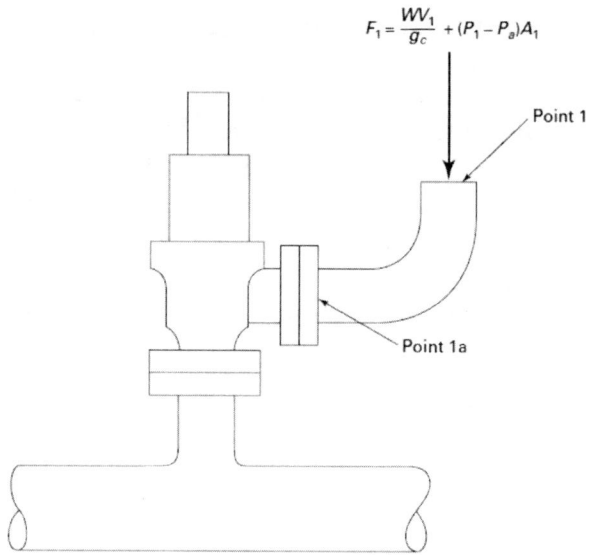

Figure 7.2A Discharge elbow (open discharge installation) [Ref. 6].

    (b) To ensure consideration of the effects of the suddenly applied load F_1, a dynamic load factor, DLF, should be applied to the piping and components (see para. 8.4).

7.2.2 Forces on vent pipes

    (a) Figure 7.2B shows the external forces resulting from a safety valve discharge, which act on the vent pipe. The vent pipe anchor and restraint system must be capable of taking the moments caused by these two forces, and also be capable of sustaining the unbalanced forces in the vertical and horizontal directions.

    (b) A bevel of the vent pipe will result in a flow that is not vertical. The equations shown are based on vertical flow. To take account for the effect of a bevel at the exit, the exit force will act at an angle, φ, with the axis of the vent pipe discharge which is a function of the bevel angle, θ. The beveled top of the vent deflects the jet approximately 30 degrees off the vertical for a 60 degree bevel, and this will introduce a horizontal component force on the vent pipe system.

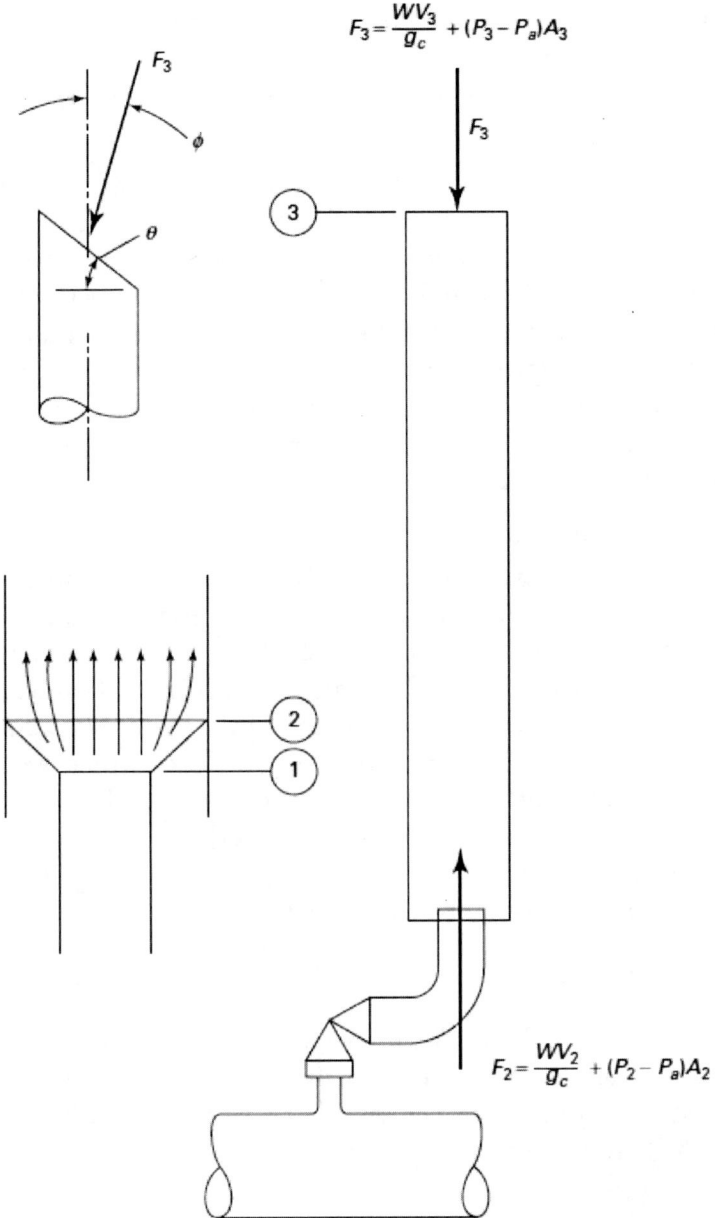

Figure 7.2B Vent pipe (open discharge installation) [Ref. 6].

8. Piping system response: loads, stresses, and reactions

8.1 Piping system loads

Dynamic loads on systems are complicated by multiple valve openings and reflected waves. The time at which these transients occur affects the system response. As different pressure waves create hoop stresses along a piping run and create bending stresses on impact with elbows and other discontinuities, stresses develop that are each damped with time. Neglecting damping, maximum stresses due to dynamic loading are described herein. The use of damping to assess system response is normally considered when performing computer analysis, and is outside the scope of this Appendix.

8.2 DLF's and dynamic pipe stresses

(a) In a piping system acted upon by time varying loads, the internal forces and moments are generally greater than those produced under static application of the load. This amplification is often expressed as the dynamic load factor, DLF, and is defined as the maximum ratio of the dynamic stress or reaction force at any time to the stress or reaction force which would have resulted from the static application of the load.

(b) For simple cases which are described by one degree of freedom models, the maximum dynamic stress (S_D) is related to the static stress (S) by the equation

$$S_D = DLF \cdot S \qquad \text{[Ref. 3] [Eq. 30]}$$

(c) The primary dynamic pipe stresses can be separated into hoop membrane stresses and longitudinal bending stresses, since their maximum values typically occur at much different times. Longitudinal membrane stresses should also be considered by the applicable Code.

(d) For static loading of a piping system, the DLF = 1.

(e) A simplified example is considered here for a 6 inch, Schedule 80, pipe and elbow. For this example, pipe supports at one end are neglected and the other end is fixed, as shown in Figure 8.2A. The resultant stresses due to the applied pressure, ΔP are shown

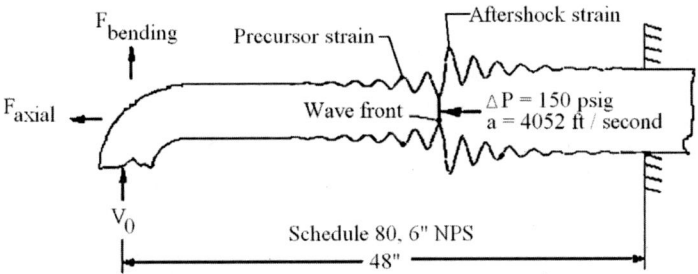

Figure 8.2A Pressure wave impingement on an elbow [Ref. 3].

in Figure 8.2B. Note that the maximum, undamped elastic stress in Figure 8.2B is twice the static bending stress at equilibrium (DLF < 2), if one assumes the loading is applied as a step change with a duration much longer than the period of the piping system. Also note that the maximum, limiting or bounding, undamped hoop stress for an elastic material is less than four times the static stress (DLF < 4) [Ref. 3], and that the hoop stresses decrease to nearly the static hoop stress at the time the bending stress reaches its maximum value. A DLF 2 < 4 is only used for fatigue and linear elastic fracture mechanics analyses.

(f) For plastic deformation during dynamic hoop stresses: 1 < DLF < 2, since damping during plastic deformation is significant and plastic precursor waves do not occur.

(g) The unbalanced forces in the pipe system need to be addressed. Axial loads, and resultant axial stresses, in piping and equipment may be significant and should be considered in the applicable pipe code equations.

(h) Experimental data may be used by the designer to establish lower DLF's.

(i) In cases where applied loads are repeated at a frequency close to a system mechanical natural frequency in such a way that an associated mode is excited, dynamic load factors on the order of 10 to 15 may be observed. This type of applied load may result from the use of some types of positive displacement pumps, which provide pulsating flows. The designer should address this design concern when fluid pulsations are induced in a piping system.

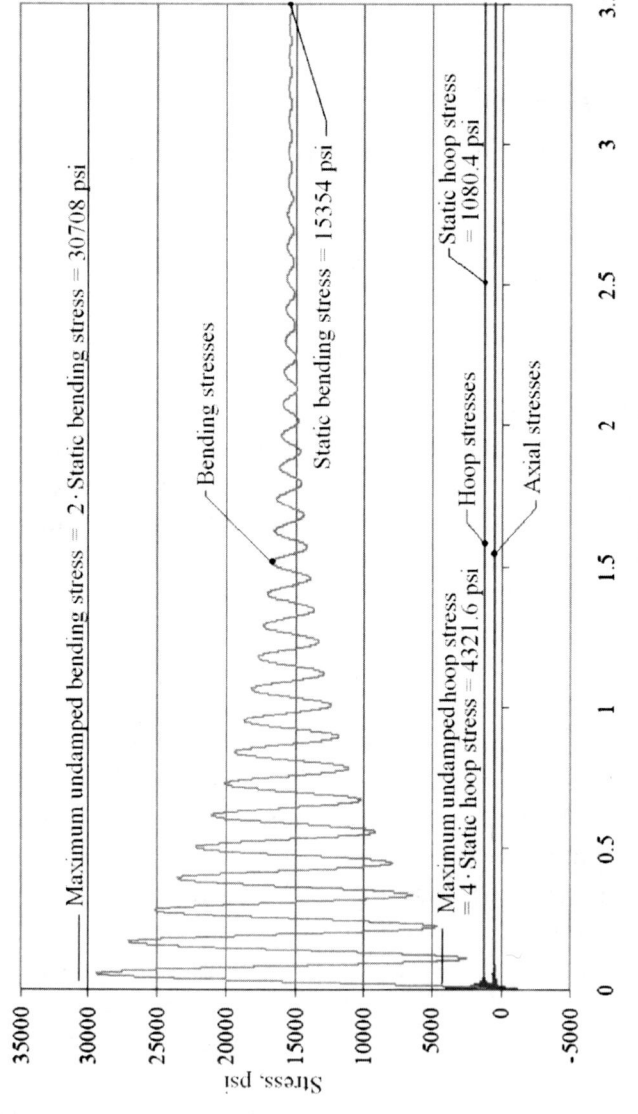

Figure 8.2B Stresses at an elbow subject to a step pressure increase [Ref. 3].

8.3 Dynamic hoop stresses

(a) Dynamic elastic hoop stresses are further complicated by the presence of precursor and aftershock vibrations as a wave travels the bore of a pipe at a nearly sonic velocity. Figure 8.2A depicts a typical hoop strain response. The maximum dynamic stress is no more than four times the static stress that would be created by a static pressure of equal magnitude to the pressure following a pressure wave.

(b) DLF's for hoop stresses vary, depending on the damage mechanism of interest. During plastic deformation, the maximum DLF varies between 1 < DLF < 2. During fatigue, elastic deformation is prevalent, and the maximum DLF varies between 2 < DLF < 4.

8.3.1 DLF's for hoop stresses

(a) For fatigue considerations, the maximum hoop stress can be expressed using a DLF < 4, where the static stress (S_{hoop}) and maximum, undamped, dynamic stress ($S_{D(hoop)}$) due to a sudden pressure increase on the inner pipe wall equal

$$S_{hoop} = \frac{\Delta P \cdot \left(OD^2 + ID^2\right)}{\left(OD^2 - ID^2\right)} \qquad \text{[Eq. 31]}$$

$$S_{D(hoop)} = DLF \cdot S < 4 \cdot \frac{\Delta P \cdot \left(OD^2 + ID^2\right)}{\left(OD^2 - ID^2\right)} \qquad \text{[Ref. 3] [Eq. 32]}$$

For the example shown in Figures 8.2A and 8.2B, the undamped dynamic stress equals four times the static stress.

(b) For fatigue considerations, the total stress (S_{total} due to a sudden pressure increase, or pressure wave, of magnitude ΔP then equals the stress due to the initial operating stress plus the dynamic effects of the pressure wave, such that

$$S_{total} < \frac{\left(P_0 + 4 \cdot \Delta P\right) \cdot \left(OD^2 + ID^2\right)}{\left(OD^2 - ID^2\right)} \qquad \text{[Ref. 3] [Eq. 33]}$$

8.3.2 Reflected pressure waves and hoop stresses

(a) Hoop stresses near the end of a closed end pipe are even higher as reflected waves double the pressure near the pipe end (see para 4.1.3). The stresses due to the incident and reflected waves add together at points along the pipe. Consequently, the highest stresses due to wave reflections occur near the closed end of a pipe, near a reducer, or near a change in pipe wall material. As a reflected pressure wave returns into the piping, the dynamic stresses due to the incident wave are also reduced due to damping.

(b) For plastic stresses due to wave reflections near the end of a closed end pipe, the maximum DLF < 4.

8.4 Dynamic bending forces and reactions

(a) Permitting simplified calculations, dynamic bending stresses and reactions due to fluid transients (such as slug flow) are caused by forces equal to

$$F(lbf) = \frac{W(lbm/s) \cdot \Delta V(ft/\sec)}{g_c(ft \cdot lbm/lbf \cdot \sec^2)} + (\Delta P(lbf/ft^2)) \cdot A(ft^2)$$

[Ref. 3] [Eq. 34]

$$F(N) = W(kg/s^2) \cdot \Delta V(m/\sec) \cdot (N \cdot \sec^2/kg \cdot m) + (\Delta P(N/m^2)) \cdot A(m^2)$$

[Ref. 3] [Eq. 35]

where W equals the mass flow rate.

(b) In liquid filled systems where a valve is suddenly closed, changes in momentum are not a significant contributor, and the forces at elbows are due primarily to pressure waves, where the resultant forces are described by

$$F(lbf) = \Delta P(lbf/ft^2) \cdot A(ft^2)$$
$$= \frac{\rho(lbm/ft^3) \cdot a(ft/\sec) \cdot \Delta V(ft/\sec) \cdot A(ft^2)}{g_c(ft \cdot lbm/lbf \cdot \sec^2)}$$

[Ref. 3] [Eq. 36]

$$F(N) = \Delta P(N/m^2) \cdot A(m^2)$$
$$= \rho(kg/m^3) \cdot (N \cdot \sec^2/kg \cdot m) \cdot a(m/\sec) \cdot \Delta V(m/\sec) \cdot A(m^2)$$
[Ref. 3] [Eq. 37]

(c) The effects of dynamic forces at elbows may be significantly affected due to the effects of similar forces at opposing elbows, i.e., U-bends or Z-bends. The distance between bends will affect the forces and resultant piping reactions due to the time of impact at each bend and the time dependent mismatch of opposing forces at the elbows.
1. As elbows are placed closer together at U-bends, forces will cancel perpendicular to the direction of the pipe runs, since the forces at the two elbows will be opposite in direction, but two distinct additive forces will occur in the direction of the pipe runs at the two elbows.
2. As elbows are placed closer together at Z-bends, forces will cancel perpendicular to the direction of the pipe runs, since the forces at the two elbows be opposite in direction, but moments will occur due to opposing forces at the two elbows in the direction of the pipe runs.

8.4.1 Bending forces and DLF's
(a) For elastic or plastic bending stresses caused by a sudden pressure increase, 1 < DLF < 2, but bending stresses may also be affected by the duration of the pressure surge or the type of loading.
(b) For example, the load may be applied as a step pressure increase, a linear ramp increase, or some other type of loading. Several types of loading are shown in Figure 8.4A, where τ equals the period of the piping vibration, and t_1 equals the rise time. Note that the step pressure increase equals the ramp response with a zero rise time.
(c) The effects of the load duration are shown in Figure 8.4B, where t_o equals the duration of the load, or in this case, the duration of the pressure surge.

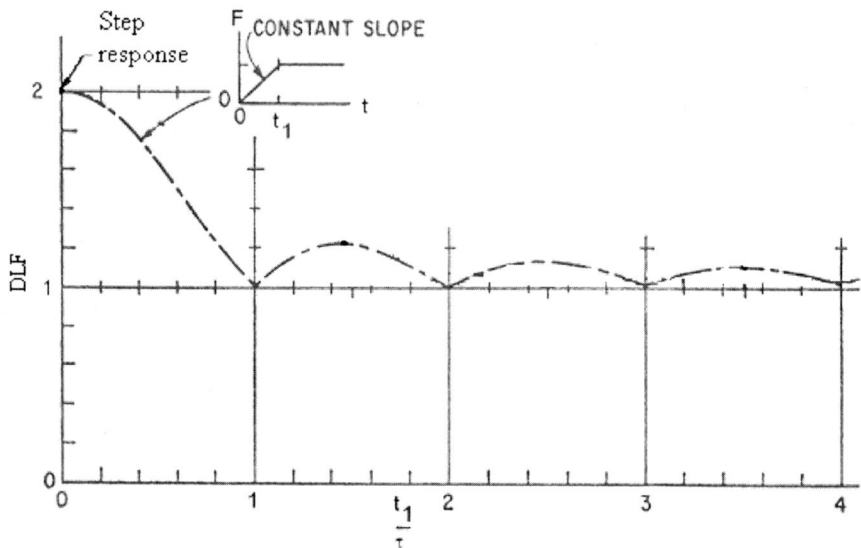

Figure 8.4A Rise time effects on loading and the DLF (Refs. 3, 7 and 213 (Harris), reprinted by permission of McGraw Hill).

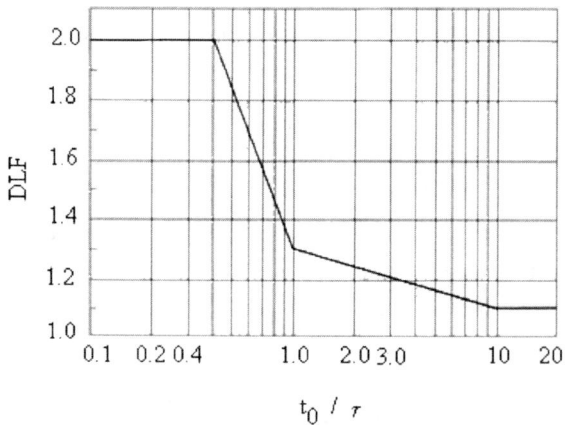

Figure 8.4B Effects of short duration pressure surges on system response (stresses or reactions) [Ref. 6].

(d) Piping configuration also significantly affects pipe stresses. If U-bends are installed, the opposing forces at the two elbows are in phase and will cancel the bending stresses for short distances between elbows. Note however that axial stresses will be doubled. As the distance between the elbows increases, the forces are increasingly unbalanced as they become out of phase with each other. However, pipe supports between elbows can reduce or even eliminate the counterbalancing forces at elbows.

(e) For Z-bends, the forces are similarly related, but a moment also occurs between the two elbows that increases as the distance between the elbows increases.

8.4.2 Reactions

(a) Reaction forces due to fluid transients are dynamic in nature. A time-history dynamic solution, incorporating a multi-degree of freedom lumped mass model solved for the transient hydraulic forces is considered to be more accurate than the simplified form of analysis presented in this Appendix.

(b) If the piping support is located at the elbow where the pressure wave impacts, then Figure 8.4B can be used to approximate the maximum value of the reaction by neglecting the reduction in reaction force due to the attached piping. Otherwise, reaction forces may be significantly reduced due to the interrelationship between the pipe vibrations and the pipe support vibrations.

(c) The effects of frequency on piping/pipe support interaction may be considered. Computer based analyses may consider the variation in vibration frequencies that accompany assumptions for pipe supports. For example, the frequency for a pipe with fixed ends varies by a factor of more than 2 when compared to that same pipe supported with simply supported ends, which have lower stiffness and yield a lower frequency.

8.5 Loads on other piping components

(a) In general, the DLF to be applied to valves is DLF = 2, since precursor strains are not expected to form in the valve body. However, in large valves there may be zones within the valves where wave reflections occur and the DLF = 4 at those locations. Fluid flow simulations may be warranted.

(b) For flange bolts, the DLF of the load applied to flange surfaces and exerted on the flange bolts is expressed as DLF < 2. However, if dynamic loads cause the bolts to stretch sufficiently to permit fluid between the gasket and the flange, additional possibly catastrophic loads may be exerted on the flange bolts.

(c) DLF's for other equipment, such as instrumentation and pressure gauges, will be similar to valve DLF's. Note also, that pressure gauges and other instrumentation frequently have inadequate response times to measure transients accurately. In other words, measured flow rates and pressures may be orders of magnitude higher than indicated if the transients are of short duration.

9. Damage assessment

(a) Damage mechanisms may occur in piping systems following valve and pump operations or condensate accumulation in steam systems. Observations include:
1. Ductile pipe failures due to valve openings and closures in liquid filled systems typically include (but are not limited to) fatigue cracks causing pipe shear and plastic deformation.
2. Valve leaks frequently precede fatigue damages during operations.
3. Fracture of brittle components.
4. Piping support damages.

(b) Corrective actions for valve operations in liquid filled systems include:
1. Automatic slow closing of valves. The definition of slow closure varies, since the closure time depends on the system resistance, where a recommended valve closure time may vary from a few seconds to a few minutes depending on pipe length. A valve closure time of $2 \cdot L/a$ does not guarantee that pressure surge magnitudes are below the maximum permitted pressures. A valve closure time of $20 \cdot L/a$ is recommended here, but faster closing times may be acceptable as determined by design.
2. Manual slow closing of valves.
3. Two speed valve closures (valve stroking) may also be used.
4. Water hammer arrestors are a smaller version of accumulator tanks, and are used to reduce wave reflection effects near the end of piping. They contain air pockets to reduce transient pressures. Arrestors are frequently used in building piping systems, when transients are created elsewhere in the supply system.
5. Note that relief valves and safety valves offer minimal control to fluid transients, since pressure waves travel past the valve into the pipe system before the valve has time to open. However, relief valves do limit the effects of reflected waves in the piping after the first wave passes the valve.

6. Stronger piping and/or additional piping restraints. The preferred alternative is to design the system operation to minimize the fluid transient.

(c) Corrective actions for transients caused by centrifugal pump operations or liquid filling of air filled pipes include:
   1. Variable speed, drive control for pump start-up and shut-down.
   2. Automatic or manual closing of pump discharge valves before starting or stopping pumps. Pump overheating and vibration should be evaluated before controlling flow using this method.
   3. Combined automatic valve and pump controls for start-up and shut-down of parallel pumps.
   4. Evaluate the effects of power losses to pump control equipment, since transients may occur due to inadvertent shut-down on loss of power to pumps.
   5. Spring operated check valves, weight operated swing check valves, or air cushioned swing check valves installed at the pump discharge.

(d) Corrective actions for vapor collapse in liquid filled systems include:
   1. Surge tanks are typically open to atmosphere and introduce liquid into the vapor space to mitigate vapor collapse.
   2. Accumulators are closed, pressurized tanks that may contain a bladder between the liquid in the pipe and the pressurizing gas in the tank. Similar to surge tank operation, accumulators also force liquid into the vapor space to mitigate vapor collapse.
   3. Vents introduce air into a vapor space to minimize the effects of vapor collapse. Air acts a cushion to the returning water column in the pipe, while vapor does not. Air effects on system performance should be considered.

(e) Common corrective actions for condensate induced water hammer include:
   1. Condensate should not be added to, or allowed to accumulate in, horizontal piping or low pipe sections of steam systems.

2. Shut down the steam system immediately if a condensate induced water hammer accident occurs. Operate valves slowly when responding to an accident to mitigate an additional transient.
3. During restart, steam should be slowly admitted into the piping system to mitigate excessive thermal stresses and condensate induced water hammer. A common technique is to admit steam through the regulator bypass valve, since the main valve typically cannot provide the low flow rates required to mitigate transients. Blow-down valves/drain valves are opened to drain water from the regulator at various points along the pipeline. When steam exits the blow-down valve, the valve is closed, and the next valve in succession drains until steam exits the piping. This process is performed for all blow-down valves in the system until condensate is removed. Then the main pressure regulator valve is opened, and the pressure regulator bypass valve is closed.
4. Steam traps should be regularly monitored to minimize condensate accumulation.
5. Safety valves and rupture discs do not open fast enough to prevent all damages from fluid transients. Typically, the first pressure wave passes the relief valve before it opens. Successive pressure waves may be reduced by these devices.
6. A common method to mitigate water cannon is to introduce air into the pipe to displace the steam prior to sudden valve closure.

(f) Fatigue damage mechanisms should be evaluated in accordance with the applicable Code. Damping values required to evaluate fatigue shall be defined by the designer.

(g) Fatigue cracks and fitness for service may be evaluated using API 579-1/ASME FFS-1.

(h) Plastic deformations and their disposition are the responsibility of the owner. Even so, plastic deformations and fatigue in piping are expected to have occurred in systems designed to existing pipe codes where transients occurred.

(i) An important observation with respect to wave velocities is that fluid transients can cause damage in pipeline systems more than a mile away from the location where the transient initiated. Failures may occur near the end of a pipe, near a corroded piping section, or may occur anywhere in the system due to the wide variation in fatigue properties for a given pipe lot. Accordingly, a fluid transient may not be immediately recognized as a cause of failure following a transient.

10. References

The following is a list of publications referenced in the development of this Appendix. The latest edition of each Appendix was referenced.

[1] Wylie, B. E., Streeter, V. L., 1993, "Fluid Transients in Systems," Prentice Hall, Upper Saddle River, New Jersey.
[2] Parmakian, J., 1963, "Waterhammer Analysis," Dover Publications, New York, New York.
[3] Leishear, R. A., 2013, "Fluid Mechanics, Water Hammer, Dynamic Stresses, and Piping Design" Publisher: ASME Press.
[4] Martin, Samuel, "Water Hammer", ASME Course Notes.
[5] Merilo, M., 1992, "Water Hammer Prevention, Mitigation, and Accommodation, Volumes 1–6, EPRI NP-6766", Stone and Webster Engineering.
[6] ASME B31.1, Power Piping.
[7] Harris, C. M., Piersol, A. G., 2002, "Harris' Shock and Vibration Handbook," McGraw Hill, New York, New York.
[8] API 579-1/ASME FFS-1, Fitness-for-Service.
[9] API Standard 521, Pressure-relieving and Depressuring Systems.
[10] API Standard 520, Sizing, Selection, and Installation of Pressure-relieving Devices in Refineries, Part I–Sizing and Selection.

11. Nomenclature

The nomenclature used herein are expressed as:

a, a_1, a_2, a_3	=	pressure wave velocities
A_1, A_2, A_3	=	areas
c	=	constant
DLF	=	dynamic load factor (dynamic magnification factor, impact factor)
E	=	modulus of elasticity
F, F_1, F_2, F_3	=	reaction force
ft	=	feet
g	=	local gravitational acceleration
g_c	=	gravitational constant
Δh	=	change in head
h_v	=	vapor pressure
h_b	=	barometric pressure
h_0	=	initial head or static head
I	=	moment of inertia
ID	=	inside pipe diameter
k	=	bulk modulus of a fluid
kg	=	kilogram
L	=	pipe length
L_s	=	slug length
L_v	=	vapor space (void) length
m	=	meter
N	=	Newton ($kg \cdot m/second^2$)
OD	=	outside pipe diameter
ΔP	=	change in pressure across a pressure wave (in the wake of a pressure wave), or change in pressure due to a change in momentum
P_a	=	atmospheric pressure
P_i	=	incident pressure wave magnitude
P_r	=	reflected pressure wave magnitude
P_t	=	transmitted pressure wave magnitude
P_{total}	=	maximum total pressure
P_1, P_2, P_3	=	steady state operating pressures
Q_1, Q_2	=	volumetric flow rates
sec	=	seconds

S	=	stress
S_D	=	dynamic stress
S_{total}	=	maximum total stress
t	=	time
t_0	=	pressure surge duration
t_1	=	rise time
$\dfrac{1}{T}$	=	pipe wall thickness
V	=	velocity
V_o	=	initial velocity
V_1, V_2, V_3	=	velocity in different pipe sections
ΔV	=	change in velocity
V_s	=	slug velocity
V_0	=	initial velocity
W	=	mass flow rate
γ	=	weight density
ρ	=	mass density
τ	=	period of vibration = 1/frequency
ν	=	Poisson's ratio

III. Updates, Revisions and Corrections to this Book

While teaching classes from this book, some discussion of uncertainty and vibration complement the other interrelated topics in this book. Discussions of "Uncertainty Analysis" and "Vibration Analysis" are followed by section on "Corrections", which includes typographical errors, misprints and clarifications.

ADDITIONAL DISCUSSION (book, p. 253)

III.A. Uncertainty Analysis

The example provided in Example 5.12 (book, Chapter 5.7) assumes some knowledge of uncertainty analysis, which is sometimes referred to as the propagation of error. A very brief discussion of a very complex subject is provided here. Errors are associated with any measurement, and there is also an error, or uncertainty, associated with different processes.

Measurement uncertainty

Measurement uncertainty is the subject of the field of metrology, which quantifies the errors associated with measuring variables, such as length, velocity, pressure, etc. (See also Coleman and Steele [184]). Sophisticated techniques have been developed to determine uncertainties and are presented in "The Evaluation of measurement data – Guide to the expression of uncertainty in measurement, JCGM 100: 2008" (Bureau International des Poids et Mesures, BIPM Pavillon de Breteuil F-92312 Sèvres Cedex, France). This document is sometimes informally referred to as the GUM, since it was based on the 1995 Guide to Uncertainty Measurement, and the term GUM persists for use in this Supplement. The GUM provides general rules to assess the errors associated with any measurement.

Errors may be distributed about the average value of a measured quantity in many forms. Errors may be distributed asymmetrically or symmetrically. They may be distributed in rectangular, triangular, sinusoidal, or some other distribution (See "Measurement Uncertainty Analysis, Principles and Methods, NASA Measurement Quality Assurance Handbook – ANNEX 3, ASA-HDBK-8739.19-3", National Aeronautics and Space Administration, Washington, DC).

The methods in the GUM predict uncertainties based on the symmetrical normal distribution shown in Figure 5.22A, which is mathematically developed from the assumption that errors occur randomly in measurement processes. To address uncertainties, the GUM uses k-values (coverage factors), which are similar to σ-values (standard deviations) discussed in statistics books. In fact, k and σ are interchangeable when using Figure 5.22A. Using the k-values, the GUM also provides methods to relate rectangular and other error distributions to the normal distribution, so that data available in different formats can be combined into uncertainties that can be expressed in terms of the normal distribution. Refer to the GUM for details. In the literature, several terms are used interchangeably to represent σ, such as RMS or as % error, as used in Example 5.12.

Other statistical terms are also used in the literature, where 2σ or $2k$ represents a commonly accepted uncertainty. In fact, ASME B89.7.3.1 (Guidelines for Decision Rules: Considering Measurement Uncertainty in Determining Conformance to Specifications) states that $2k$ should be used in uncertainty calculations. This $2k$ uncertainty is frequently referred to as an uncertainty of 95% confidence, but the confidence level for $2k$ is more precisely 95.45% as shown in Table 5.1.

To use Table 5.1, one needs to know the number of experimental, or measured, data points. Then $v = n - 1$, where n equals the number of data points. Reading v from the left side of Table 5.1, the desired uncertainty may be obtained by reading the appropriate column. For example, if an infinite number of values are available and a $2k$ uncertainty is desired, the percentage, p, or confidence level equals 95.45%. That is, 19 times out of twenty the calculation will correctly predict new data points to lie within the range of $\pm 2k$.

An example is warranted to clarify these terms. Return to Example 5.12, where a k value was approximated at 6.6%, assuming that uncertainties were known to at least 95% confidence. Then, for a calculated fluid transient pressure of 250 psig, the pressure can be expressed as $250 \pm 2 \cdot 0.066 \cdot 250 = 250 \pm 33$ psig, and any pressure measurements in a system are expected to occur within this range of pressure approximately nineteen times out of twenty. Note also from Table 5.1 that $3k$ and an infinite number of data points provide a 99.73% confidence level.

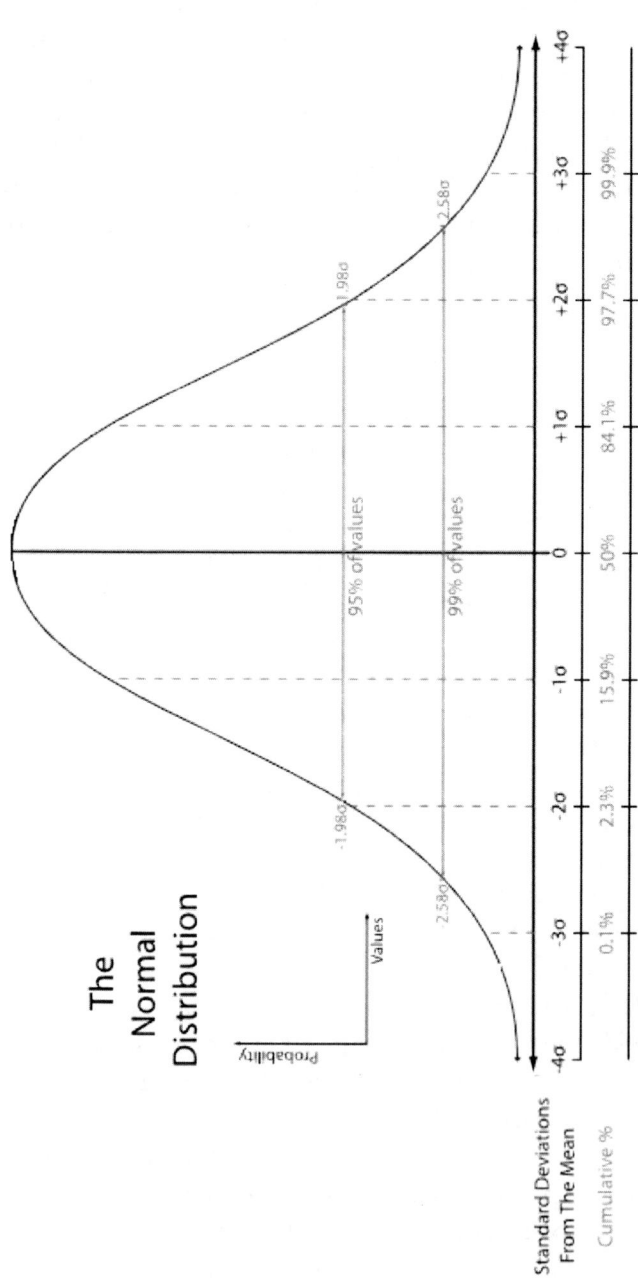

Figure 5.22A Uncertainty associated with the normal distribution (Wikipedia).

Table 5.1 Student-T values (Wikipedia).

Degrees of Freedom ν	Fraction p in Percent					
	68.27	90	95	95.45[a]	99	99.73[a]
1	1.84	6.31	12.71	13.97	63.66	235.80
2	1.32	2.92	4.30	4.53	9.92	19.21
3	1.20	2.35	3.18	3.31	5.84	9.22
4	1.14	2.13	2.78	2.87	4.60	6.62
5	1.11	2.02	2.57	2.65	4.03	5.51
6	1.09	1.94	2.45	2.52	3.71	4.90
7	1.08	1.89	2.36	2.43	3.50	4.53
8	1.07	1.86	2.31	2.37	3.36	4.28
9	1.06	1.83	2.26	2.32	3.25	4.09
10	1.05	1.81	2.23	2.28	3.17	3.96
11	1.05	1.80	2.20	2.25	3.11	3.85
12	1.04	1.78	2.18	2.23	3.05	3.76
13	1.04	1.77	2.16	2.21	3.01	3.69
14	1.04	1.76	2.14	2.20	2.98	3.64
15	1.03	1.75	2.13	2.18	2.95	3.59
16	1.03	1.75	2.12	2.17	2.92	3.54
17	1.03	1.74	2.11	2.16	2.90	3.51
18	1.03	1.73	2.10	2.15	2.88	3.48
19	1.03	1.73	2.09	2.14	2.86	3.45
20	1.03	1.72	2.09	2.13	2.85	3.42
25	1.02	1.71	2.06	2.11	2.79	3.33
30	1.02	1.70	2.04	2.09	2.75	3.27
35	1.01	1.70	2.03	2.07	2.72	3.23
40	1.01	1.68	2.02	2.06	2.70	3.20
45	1.01	1.68	2.01	2.06	2.69	3.18
50	1.01	1.68	2.01	2.05	2.68	3.16
100	1.005	1.660	1.984	2.025	2.626	3.077
∞	1.000	1.645	1.960	2.000	2.576	3.000

The student-T values presented in Table 5.1 also provide a means to evaluate limited experimental data. For example, if only four values are available from a series of tests, the student-T equals 3.31 for a 95% confidence level. If only two data points are available, $k = 13.97$. For a 95% confidence level, $13.97k$ describes the variation about the mean value of the two data points. In other words, the uncertainty increases significantly when few data points are available for analysis.

Process variations

Frequently, process uncertainties are required with respect to a mathematical curve. Regression analysis is an uncertainty calculation with respect to a line, either curved or straight. Although outside the scope of this text, the technique typically uses a method referred to as a Gaussian distribution to determine a curve by calculating the square root of the sum of the squares of the experimentally determined values of numerous data points. For example, Equations 7.92 and 7.93, which describe damping, were determined by using linear regression analysis with respect to many data points obtained from measured piping vibrations in steam piping systems.

ADDITIONAL DISCUSSION (p. 324)
III.B. Vibration Analysis

The field of vibration analysis focuses on experimental measurements of vibrations and their interpretation with respect to equipment performance. A brief discussion is presented here to relate field vibration measurements to piping systems. Although any system can be analyzed using these techniques, rotational equipment like motors and pumps are typical candidate for analysis. Fast Fourier transforms are used to convert raw vibration data into discrete vibrations, as depicted in Figure 7.16. Once these vibrations are identified, they can be analyzed to determine the causes of vibrations and possible resulting failures. The vibrations must first be identified, and may then be evaluated for their effects. The first objective then is to identify the causes of excessive vibrations.

The overall vibration is referred to as an unfiltered vibration, and the discrete vibrations are referred to as filtered vibrations. Two types of vibrations are commonly observed, i.e., resonant modal frequencies and forcing frequencies. If the forcing frequency is nearly coincident to any modal frequency of a system, the system will vibrate at that frequency.

For example, if a pipe is attached to a pump, it will vibrate excessively in resonance if the motor speed is coincident to any one of the natural frequencies of the piping.

Ball bearing damages present another example of resonant frequencies. A ball bearing consists of multiple components, which include the balls, the inner and outer races on which the balls move, and the ball cages that separate the balls. Each of these components has manufacturer identified natural frequencies. When one, or several, of these components is damaged, vibrations occur at their respective natural frequencies, where failures typically yield audible high pitched squealing sounds when races are damaged or incipient tapping sounds heard with a stethoscope when balls are damaged.

For forcing frequencies, the measured vibration occurs at multiples of the pump or motor operating speed, where the vibrations are related to geometry. If a pump shaft is bent or misaligned, the vibration will be one or two times the operating speed. If a three jaw coupling is damaged, the vibration will be three times the operating speed. If a 52 tooth coupling is damaged, the vibration will be 52 times the operating speed. If cavitation is present on pump impeller blades, the vibration will equal the number of impellers times the operating speed. Vertical pump seals fail at one half of the operating speed due to fluid whirl in journal bearings. Pump seals in horizontal pup frequently fail due to overheating or resonant vibrations of the pump and piping. In other words, the skill in vibration analysis is to view the vibration spectrum provided by the vibration analyzer screen, and determine which component of the system is causing the primary vibrations.

Once the vibrations are identified, they need to be evaluated for acceptability. A commonly used set of figures to evaluate vibrations are presented as Figures 7.22A–7.22C, where vibrations are expressed in terms of displacement, velocity, or acceleration. These figures are used as screening criteria to determine if measured velocities are acceptable. Some prudence should be used when applying these figures, since all vibration factors are not considered. For example, if a motor is cantilevered, much higher acceptable vibrations may be observed. Also, if vibrations exceed the values shown, and equipment has operated for years without failure, caution should be exercised to prevent trying to fix a problem that does not exist.

Note that the vibration velocity is slightly above 0.1 inches per second for operating speeds of 1800 to 3600 rpm. This observation is widely

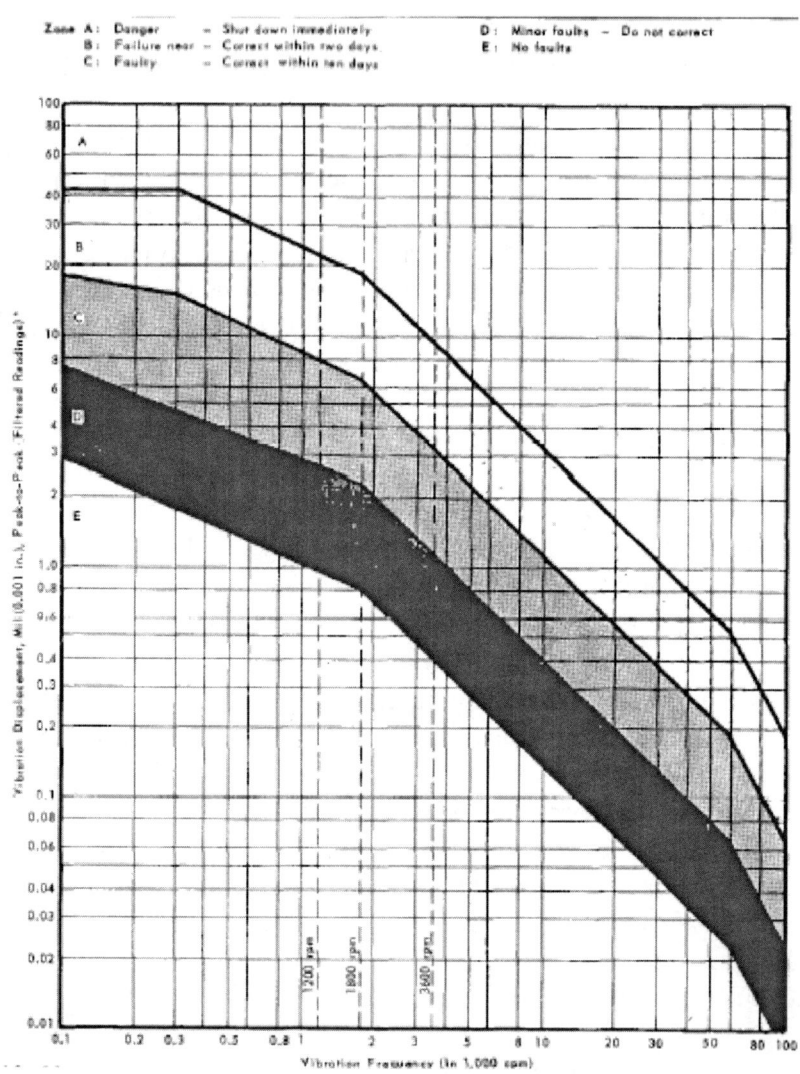

Figure 7.22A Filtered vibration, displacement.

used to claim that vibration level below 0.1 inches per second need not be considered further in any damage assessment. In fact, this value is commonly used as a first approximation to assess damages for piping near pumps and compressors.

To measure vibrations, various vibration analyzers are available. Small hand held instruments measure the maximum unfiltered vibration for quick results, requiring little skill to take measurements. Larger, more

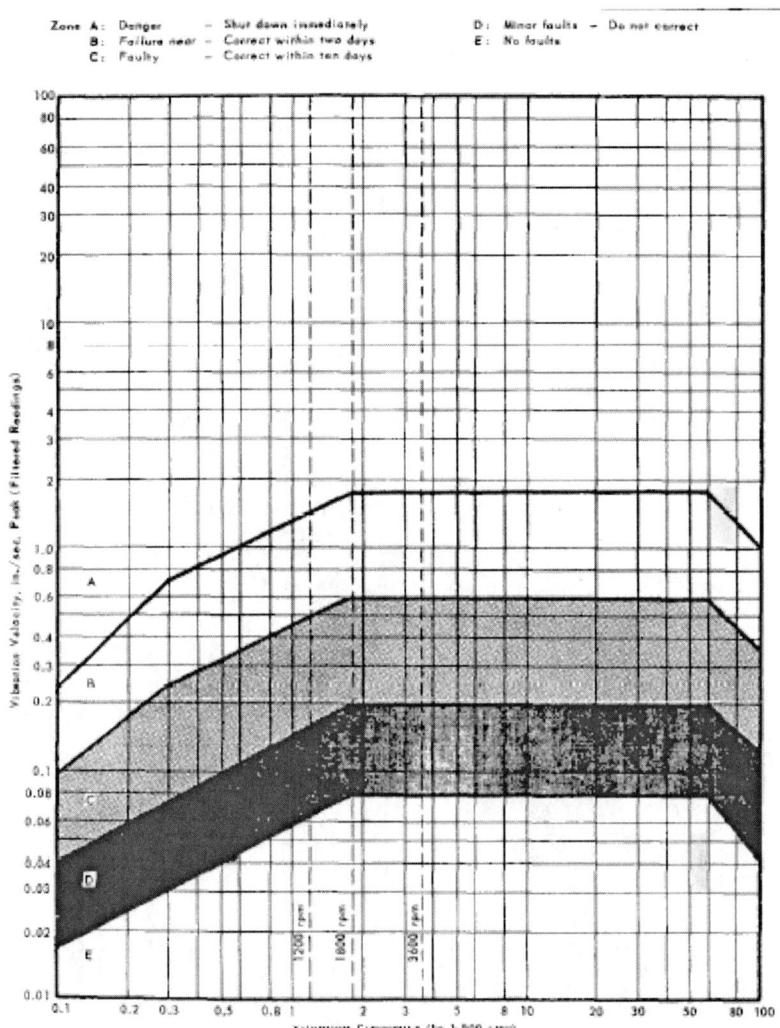

Figure 7.22B Filtered vibration, velocity.

expensive equipment, provided with a shoulder strap, can be used in the field to take vibration measurements and read unfiltered vibration spectrums immediately. Some of these units can also be used to download vibration spectrums for automatic comparison to tabular manufacturer's data for bearing and belt frequencies.

With respect to fluid transients, vibration analyzers typically will not measure high frequency hoop stresses accurately, but will measure

108 Updates, revisions and corrections

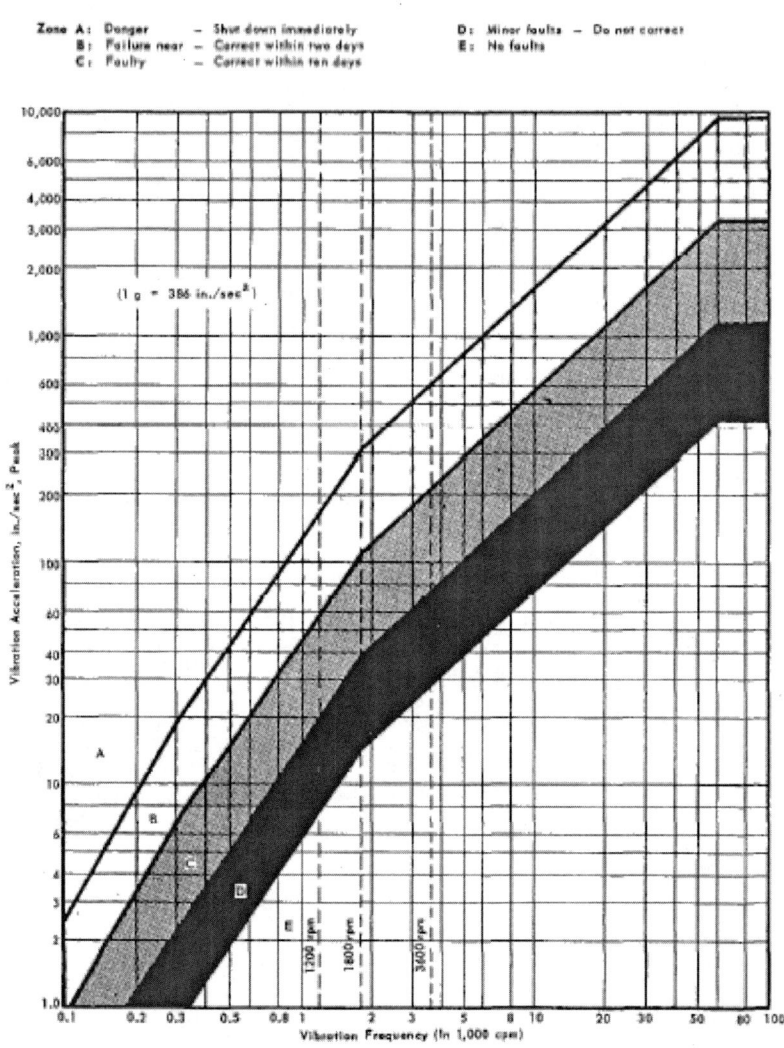

Figure 7.22C Filtered vibration, acceleration.

bending stresses. Even so, strain gauges can be selected to provide more accurate vibration measurements during fluid transients. However, strain gauges require permanent attachment to the piping, while vibration analyzers simply require a probe to be placed against the piping to obtain measurements. A comparison of vibration results from strain gauges versus vibration analyzers has not been fully investigated.

III.C. Corrections

REVISE (book, throughout, modification) "DMF" to "DLF"

The Mechanical Design Committee (MDC) for the B31 Piping Codes prefers the abbreviation DLF (dynamic load factor), which is recommended in lieu of DMF (Dynamic magnification factor) and is used in ASME, B31.1. Accordingly, DLF is expected to become the consistently used abbreviation in the piping industry, rather than DMF which is used extensively throughout this book.

REVISE (book, p. 3, modification)
1.2.2
FROM: "dynamic magnification factor"
TO: "dynamic load factor"

REVISE (book, p. 33, clarification)
Figure 2.19
FROM: "REYNOLDS NUMBER R"
TO: "REYNOLDS NUMBER Re"

REVISE (book, p. 39)

FROM: $K_3 = K_2 \cdot \left(\dfrac{D_3}{D_2}\right)^4 = K_2 \cdot \left(\dfrac{2.067}{3.068}\right)^4 = K_2 \cdot 0.206$

TO: $K_3 = K_2 \cdot \left(\dfrac{D_2}{D_3}\right)^4 = K_2 \cdot \left(\dfrac{2.067}{3.068}\right)^4 = K_2 \cdot 0.206$

REVISE (book, p. 75)
FROM: "Linear valves change the flow rate linearly as the valve stem position changes, and equal percentage valves change the flow exponentially with respect to the change in valve stem position."

REVISE (book, p. 30)
FROM:

Liquid	*Sp Gr at 60 F	VISCOSITY SSU	VISCOSITY Centistokes	At F
Freon	1.37 to 1.49 @ 70 F		.27–.32	70
Glycerine (100%)	1.26 @ 68 F	2,950 / 813	648 / 176	68.6 / 100
Glycol:				
Propylene	1.038 @ 68 F	240.6	52	70
Triethylene	1.125 @ 68 F	185.7	40	70
Diethylene	1.12	149.7	32	70
Ethylene	1.125	88.4	17.8	70
Hydrochloric Acid (31.5%)	1.05 @ 68 F		1.9	68
Mercury	13.6		.118 / .11	70 / 100
Phenol (Carbolic Acid)	.95 to 1.08	65	11.7	65
Silicate of Soda	40 Baume / 42 Baume	365 / 637.6	79 / 138	100 / 100
Sulfuric Acid (100%)	1.83	75.7	14.6	68
FISH AND ANIMAL OILS:				
Bone Oil	.918	220 / 65	47.5 / 11.6	130 / 212
Cod Oil	.928	150 / 95	32.1 / 19.4	100 / 130
Lard	.96	287 / 160	62.1 / 34.3	100 / 130
Lard Oil	.912 to .925	190 to 220 / 112 to 128	41 to 47.5 / 23.4 to 27.1	100 / 130
Menhaden Oil	.933	140 / 90	29.8 / 18.2	100 / 130
Neatsfoot Oil	.917	230 / 130	49.7 / 27.5	100 / 130
Sperm Oil	.883	110 / 78	23.0 / 15.2	100 / 130
Whale Oil	.925	163 to 184 / 97 to 112	35 to 39.6 / 19.9 to 23.4	100 / 130
MINERAL OILS:				
Automobile Crankcase Oils (Average Midcontinent Paraffin Base):				
SAE 10	**.880 to .935	165 to 240 / 90 to 120	35.4 to 51.9 / 18.2 to 25.3	100 / 130
SAE 20	**.880 to .935	240 to 400 / 120 to 185	51.9 to 86.6 / 25.3 to 39.9	100 / 130
SAE 30	**.880 to .935	400 to 580 / 185 to 255	86.6 to 125.5 / 39.9 to 55.1	100 / 130
SAE 40	**.880 to .935	580 to 950 / 255 to 80	125.5 to 205.6 / 55.1 to 15.6	100 / 130 / 210
SAE 50	**.880 to .935	950 to 1,600 / 80 to 105	205.6 to 352 / 15.6 to 21.6	100 / 210
SAE 60	**.880 to .935	1,600 to 2,300 / 105 to 125	352 to 507 / 21.6 to 26.2	100 / 210
SAE 70	**.880 to .935	2,300 to 3,100 / 125 to 150	507 to 682 / 26.2 to 31.8	100 / 210
SAE 10W	**.880 to .935	5,000 to 10,000	1,100 to 2,200	0
SAE 20W	**.880 to .935	10,000 to 40,000	2,200 to 8,800	0
Automobile Transmission Lubricants:				
SAE 80	**.880 to .935	100,000 max	22,000 max	0
SAE 90	**.880 to .935	800 to 1,500 / 300 to 500	173.2 to 324.7 / 64.5 to 108.2	100 / 130
SAE 140	**.880 to .935	950 to 2,300 / 120 to 200	205.6 to 507 / 25.1 to 42.9	130 / 210
SAE 250	**.880 to .935	Over 2,300 / Over 200	Over 507 / Over 42.9	130 / 210
Crude Oils:				
Texas, Oklahoma	.81 to .916	40 to 783 / 34.2 to 210	4.28 to 169.5 / 2.45 to 45.3	60 / 100
Wyoming, Montana	.86 to .88	74 to 1,215 / 46 to 320	14.1 to 263 / 6.16 to 69.3	60 / 100
California	.78 to .92	40 to 4,840 / 34 to 700	4.28 to 1,063 / 2.4 to 151.5	60 / 100
Pennsylvania	.8 to .85	46 to 216 / 38 to 86	6.16 to 46.7 / 3.64 to 17.2	60 / 100
Diesel Engine Lubricating Oils (Based on Average Midcontinent Paraffin Base): Federal Specification No. 9110	**.880 to .935	165 to 240 / 90 to 120	35.4 to 51.9 / 18.2 to 25.3	100 / 130

*Unless otherwise noted. **Depends on origin or percent and type of solvent.

UPDATES, REVISIONS AND CORRECTIONS

TO:

Liquid	*Sp Gr at 60 F	VISCOSITY SSU	VISCOSITY Centistokes	At F
Diesel Engine Lubricating Oils (Based on Average Midcontinent Paraffin Base):				
Federal Specification No. 9170	**.880 to .935	300 to 410 / 140 to 180	64.5 to 88.8 / 29.8 to 38.8	100 / 130
Federal Specification No. 9250	**.880 to .935	470 to 590 / 200 to 255	101.8 to 127.8 / 43.2 to 55.1	100 / 130
Federal Specification No. 9370	**.880 to .935	800 to 1,100 / 320 to 430	173.2 to 238.1 / 69.3 to 93.1	100 / 130
Federal Specification No. 9500	**.880 to .935	490 to 600 / 92 to 105	106.1 to 129.9 / 18.54 to 21.6	130 / 210
Diesel Fuel Oils:				
No. 2 D	**.82 to .95	32.6 to 45.5 / 39	2 to 6 / 1 to 3.97	100 / 130
No. 3 D	**.82 to .95	45.5 to 65 / 39 to 48	6 to 11.75 / 3.97 to 6.78	100 / 130
No. 4 D	**.82 to .95	140 max / 70 max	29.8 max / 13.1 max	100 / 130
No. 5 D	**.82 to .95	400 max / 165 max	86.6 max / 35.2 max	122 / 160
Fuel Oils:				
No. 1	**.82 to .95	34 to 40 / 32 to 35	2.39 to 4.28 / 2.69	70 / 100
No. 2	**.82 to .95	36 to 50 / 33 to 40	3.0 to 7.4 / 2.11 to 4.28	70 / 100
No. 3	**.82 to .95	35 to 45 / 32.8 to 39	2.69 to .584 / 2.06 to 3.97	100 / 130
No. 5A	**.82 to .95	50 to 125 / 42 to 72	7.4 to 26.4 / 4.91 to 13.73	100 / 130
No. 5B	**.82 to .95	125 to 400 / 72 to 310	26.4 to 86.6 / 13.63 to 67.1	100 / 122 / 130
No. 6	**.82 to .95	450 to 3,000 / 175 to 780	97.4 to 660 / 37.5 to 172	122 / 160
Fuel Oil—Navy Specification	**.989 max	110 to 225 / 63 to 115	23 to 48.6 / 11.08 to 23.9	122 / 160
Fuel Oil—Navy II	1.0 max	1,500 max / 480 max	324.7 max / 104 max	122 / 160
Gasoline	.68 to .74		.46 to .88 / .40 to .71	60 / 100
Gasoline (Natural)	76.5 degrees API		.41	68
Gas Oil	28 degrees API	73 / 50	13.9 / 7.4	70 / 100
Insulating Oil: Transformer, switches and circuit breakers		115 max / 65 max	24.1 max / 11.75 max	70 / 100
Kerosene	.78 to .82	35 / 32.6	2.69 / 2	68 / 100
Machine Lubricating Oil (Average Pennsylvania Paraffin Base):				
Federal Specification No. 8	**.880 to .935	112 to 160 / 70 to 90	23.4 to 34.3 / 13.1 to 18.2	100 / 130
Federal Specification No. 10	**.880 to .935	160 to 235 / 90 to 120	34.3 to 50.8 / 18.2 to 25.3	100 / 130
Federal Specification No. 20	**.880 to .935	235 to 385 / 120 to 185	50.8 to 83.4 / 25.3 to 39.9	100 / 130
Federal Specification No. 30	**.880 to .935	385 to 550 / 185 to 255	83.4 to 119 / 39.9 to 55.1	100 / 130
Mineral Lard Cutting Oil:				
Federal Specification Grade 1		140 to 190 / 86 to 110	29.8 to 41 / 17.22 to 23	100 / 130
Federal Specification Grade 2		190 to 220 / 110 to 125	41 to 47.5 / 23 to 26.4	100 / 130
Petrolatum	.825	100 / 77	20.6 / 14.8	130 / 160
Turbine Lubricating Oil: Federal Specification (Penn Base)	.91 Average	400 to 440 / 185 to 205	86.6 to 95.2 / 39.9 to 44.3	100 / 130
VEGETABLE OILS:				
Castor Oil	.96 @ 68 F	1,200 to 1,500 / 450 to 600	259.8 to 324.7 / 97.4 to 129.9	100 / 130
China Wood Oil	.943	1,425 / 580	308.5 / 125.5	69 / 100
Cocoanut Oil	.925	140 to 148 / 76 to 80	29.8 to 31.6 / 14.69 to 15.7	100 / 130
Corn Oil	.924	135 / 54	28.7 / 8.59	130 / 212
Cotton Seed Oil	.88 to .925	176 / 100	37.9 / 20.6	100 / 130

*Unless otherwise noted. **Depends on origin or percent and type of solvent.

TO: "Linear valves change the flow rate linearly as the valve stem position changes, and equal percentage valves change the flow exponentially with respect to a linear change in valve stem position."

REVISE (book, p. 100, error)
FROM:
"Inertial data should be obtained from motor manufacturers, but data for numerous pumps has been established (Thorley [68]), according to

$$I_p = 0.2435 \cdot \left(\frac{P'}{n^3}\right)^{0.9556} \qquad \text{US [Eq. 2.101]}$$

$$I_m = 0.0043 \cdot \left(\frac{P'}{n^3}\right)^{1.48} \qquad \text{US [Eq. 2.102]}$$

To find the rated power required for Equations 2.101 and 2.102, rewrite Equation 2.89, such that
....
... I_p is the pump rotary inertia in lbf/ft^2, I_m is the motor rotary inertia, ..."
TO:
"Inertial data should be obtained from motor manufacturers, but data for numerous lightweight pumps has been established by Thorley [68], according to

$$I_p = 5.74 \cdot 10^{-12} \cdot \left(\frac{P'}{n^3}\right)^{0.9556} \qquad \text{US [Eq. 2.101]}$$

$$I_m = 1.00 \cdot 10^{-8} \cdot \left(\frac{P'}{n}\right)^{1.48} \qquad \text{US [Eq. 2.102]}$$

To find the rated power at maximum efficiency required for Equations 2.101 and 2.102, rewrite Equation 2.89, such that
....
... I_p is the pump rotary inertia in lbm-ft^2, I_m is the motor rotary inertia, ..."

REVISE (book, p. 121, Table 3.1, error)
FROM: n
TO: n'.

REVISE (book, p. 128, Figure 3.8, error)
FROM: N'
TO: n'.

REVISE (book, p. 161, new information)
FROM: The stress range factor, f'', accounts for fatigue and is obtained from Figure 3.31.
TO: The stress range factor, f'', accounts for fatigue and is obtained from Figure 3.31. In-process changes to ASME B31.3 provide additional data on fatigue as shown in Figure 3.33A. Note that a fatigue limit does not exist when the structural stress method is used, where weld flaws change fatigue performance from that expected from fatigue tests performed with test specimens having smooth surfaces.

REVISE (book, p. 164, error)
FROM: and the Code fatigue curve was generated for $R' = 0$.
TO: and the Code fatigue curve was generated for $R' = -1$.

REVISE (book, p. 168, additions)
FROM: Other piping design references include Peng and Peng [135], Kannapann [136], and Grinnel [137].
TO: Other piping design references include Peng and Peng [135], Kannapann [136], Grinnel [137], Rodabaugh, E. C., Terao, D. 1992, "Functional Capability of Piping Systems", NUREG-1367, and Burgreen, D., 1975, "Design Methods for Power Plant Structures", Arcturus Publishers, Cherry Hills, New Jersey.

REVISE (book, p. 215, error)
FROM: Using the B31.3 fatigue curve (Figure 3.32),
TO: Using the B31.3 fatigue curve (Figure 3.31),
FROM: Using the pressure vessel fatigue curve (Figure 3.32),
TO: Using the pressure vessel fatigue curve (Figure 3.33),

114 Updates, revisions and corrections

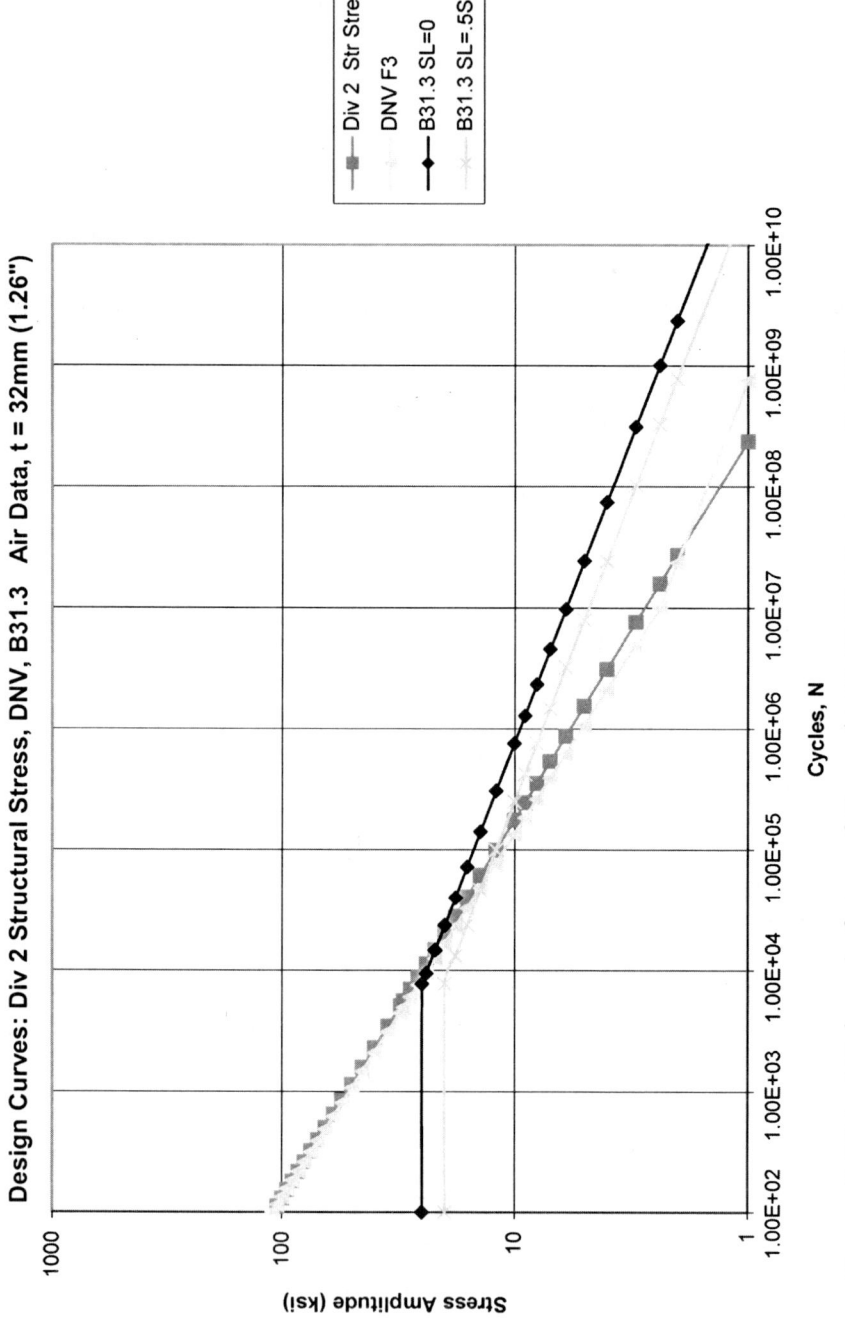

Figure 3.33A High cycle fatigue failures (ASME B31.3, Appendix W).

REVISE (book, p. 217, additions)
FROM: An actual system operated under these conditions for 150,000 to 300,000 stress cycles without fracture indications, which implies that the fatigue limit is higher than the value obtained from the pressure vessel fatigue curve.
TO: An actual system operated under these conditions for 150,000 to 300,000 stress cycles without fracture indications, which implies that the fatigue limit may be higher than the value obtained from the pressure vessel fatigue curve. However, the number of cycles may have also been less than estimated, where observations showed that the piping vibrations decreased several weeks after system start-up, and the number of start-ups was not recorded.

REVISE (book, p. 233, additions)
FROM: Wiggert and Tijsseling [179] also provided a summary of theory with numerous references.
TO: Wiggert and Tijsseling [179] also provided a summary of theory with numerous references. Additional information is provided along with commercial software, such as Whamo, Relap, Wanda, Flowmaster®, Impulse, Hammer, BOS Fluids®, Bentley Nevada®, Autopipe®, and Pipenet. Additionally, a comprehensive discussion of water hammer equations and theory was provided in "A Review of Water Hammer Theory and Practice", M. Ghidaoui, M. Zhao, D. McInnis, and D. Axworthy, ASME, Applied Mechanics Reviews, 58, 2005. The energy Institute also provides "Guidelines for the Avoidance of Vibration Induced Fatigue Failure in Process Pipework".

REVISE (book, p. 312, modifications)
FROM: The dynamic magnification factor, DMF, has several equivalent terms in the literature and is referred to as the dynamic load factor, maxi-max response, impact factor, percent overshoot plus 1 (P.O. + 1), transmissibility, and dynamic amplification factor. The term DMF is being used in this work for consistency, since DMF has gained some acceptance in the piping industry.
TO: The dynamic load factor, DLF, has several equivalent terms in the literature and is referred to as the dynamic magnification factor, maxi-max response, impact factor, percent overshoot plus 1 (P.O. + 1),

transmissibility, and dynamic amplification factor. The term DLF is being used in this work for consistency, since DLF has gained some acceptance in the piping industry.

REVISE (book, p. 351, error)
FROM: "thick wall, 2-in. ID aluminum tube."
TO: "thick wall, 6-in. ID aluminum tube."

REVISE (book, p. 350 and 351, error)
Equations 8.91 and 8.92 should be shown as

$$w_a = \frac{P \cdot r^2}{E \cdot \overline{T}} \cdot c \cdot \left(1 - 3 \cdot {}^{-d \cdot \xi} \cdot e^{-\xi} \cdot \left(\cos(c \cdot \xi)\right)\right) \qquad \text{US [Eq. 8.91]}$$

$$w_{pr} = -2 \cdot \frac{P \cdot r^2}{E \cdot \overline{T}} \cdot e^{d \cdot \xi} \cdot e^{\xi} \cdot \left(\cos(c \cdot \xi)\right) \qquad \text{US [Eq. 8.92]}$$

REVISE (book, p. 373, revision)

FROM: Theoretically, water hammer may have compressed the trapped hydrogen gas to cause auto-ignition of the hydrogen/oxygen mixture in the pipe. Given sufficient gas accumulation in the pipe due to radiolytic decomposition of water, a flame front to explosion transition may have occurred to burst the pipe. Research is required to adequately demonstrate either theory.
TO: An article in the December, 2014, edition of the ASME Mechanical Engineering Magazine, demonstrated that water hammer compressed hydrogen and oxygen to ignite a hydrogen fire, where "The spark that ignited Three Mile Island burst from a safety valve", by R. A. Leishear. The article summarized the sequence of events for this near explosion of a nuclear reactor building on the day of a nuclear reactor meltdown. Additional discussion for the interactions between water hammer and pipeline explosions were presented by the Mensa World Journal in an article titled, "Explosions: a fresh look at Chernobyl, Three Mile Island, the Gulf Oil Spill and Fukushima Daiichi", by R. A. Leishear."

REVISE (book, p. 388, error)
FROM: "…=1225.2   US (9.18)"
TO: "…< 2·1225.2   US (9.18)"

Index

A
Air effects on pressure surges, 69–71
Autofrettage, 48

B
Bending stresses, 15–18, 33–34, 53, 85–86, 88–91
Brittle failure, 13, 93

C
Centrifugal pump operations, 48, 66–68, 94
Condensate induced water hammer, 51, 75–80, 94–95
Critical velocity, 31

D
Damping, 42
Dead head pressure, 48
Design pressure, 48
DLF, 12, 15, 16, 25, 30–32, 39, 48, 58, 84, 87, 89, 91–92, 109
DMF, *See* DLF
Ductile failure, 76, 78–79, 93
Dynamic load factor, *See* DLF
Dynamic stress theory, 25, 116

E
Explosions in piping, 116

F
Fatigue, 13, 48, 87, 113–114
Flanges, 91
Flexural resonance, 25
Frequency, 30, 91

H
Head, 48
Hoop stresses, 15, 21–32, 34–37, 85, 87–88
Hydraulic grade line, 49, 73

I
Inertia, pumps and motors, 112

L
Line pack, 57

N
Normal distribution, 102

O
Operating pressure, 49

P
Plastic deformation, 13, 15, 38–43, 78, 85, 95
Positive displacement pumps, 49, 68
Pressure vessel response, 19
Pressure wave, 51, *See* Water hammer equation

R
Ratcheting, 49
Reaction forces, 81–83, 91
Resonance, 54, 85
Run–out, 49

S
SDOF vibrations, 16, 17, 23, 33
Single degree of freedom, *See* SDOF
Slug flow, 51, 74, 79–80

Strain rate, 50
Student-T, 103

U
Uncertainty analysis, 100–104

V
Valve closure time, 54, 93
Valve leaks, 93
Valve response, 19–22, 91
Vapor collapse, 72–76, 94

Vibration analysis, 104
Viscosities, 110–111
Void formation, *See* Vapor collapse

W
Water cannon, 77, 95
Water hammer equation, 55–59
Wave speed, 55–57
Wave speeds, 31, 55, 59–62, 96
Waves, Reflected, 13, 15, 36, 62–66, 88